MAGNESIUM, MAGNESIUM ALLOYS, AND MAGNESIUM COMPOSITES

MAGNESIUM, MAGNESIUM ALLOYS, AND MAGNESIUM COMPOSITES

Manoj Gupta
Nai Mui Ling Sharon

WILEY

A JOHN WILEY & SONS, INC. PUBLICATION

Published by John Wiley & Sons, Inc., Hoboken, New Jersey.
Published simultaneously in Canada

For general information on our other products and services or for technical support, please
contact our Customer Care Department within the United States at (800) 762-2974, outside the
United States at (317) 572-3993 or fax (317) 572-4002.

Wiley also publishes its books in a variety of electronic formats. Some content that appears in
print may not be available in electronic formats. For more information about Wiley products,
visit our web site at www.wiley.com.

Library of Congress Cataloging-in-Publication Data:

Magnesium, magnesium alloys, and magnesium composites / authored by
Manoj Gupta, Nai Mui Ling Sharon.
    p. cm.
  Includes index.
  ISBN 978-0-470-49417-2 (hardback)
  1. Magnesium.   2. Magnesium compounds.   3. Magnesium alloys.
  4. Metallic composites.   I. Gupta, M. (Manoj)   II. Nai, Mui Ling Sharon.
  TA480.M3M325 2011
  620.1′87–dc22

                                      2010018479

Printed in the United States of America

10 9 8 7 6 5 4 3 2 1

CONTENTS

5 MAGNESIUM COMPOSITES   113

PREFACE

Rapidly depleting energy sources and an alarming increase in emission of green house gases has been catalytic in driving scientists and researchers all over the world to find ways to utilize energy efficiently. One possible solution to cut down on energy consumption is to replace materials with inherently high density with those that have comparatively low density. Among the metallic materials, magnesium-based materials provide the greatest potential to minimize energy consumption in a diverse range of engineering applications.

Magnesium is the sixth most abundant element in the earth crust and the third most abundant dissolved mineral in the seawater. Magnesium is also the lightest of all structural metals. It has a density of 1.74 g/cm^3, which is about one-fourth the density of steel, and about two-thirds that of aluminum. Due to its low density and high specific mechanical properties, magnesium-based materials are popular materials for applications in the automotive, aerospace, electronics, and sports equipment industries.

There has been a limited number of books on magnesium-based materials and the technology of magnesium. Although the *ASM Handbook* provided an extensive collection of the properties of magnesium alloys, there has been no attempt to consolidate and summarize the properties of other futuristic magnesium-based materials, in particular magnesium-based composites.

Accordingly, this book aims to provide readers with an insight into the science, characteristics, and applications of current and futuristic magnesium-based materials. Particular emphasis is placed on the properties of magnesium-based composites and the effects of reinforcements on the properties of the resultant composites.

This book is targeted as a reference book for engineers, scientists, technicians, teachers, and students in the fields of materials design, development and selection, manufacturing, and engineering.

Sharon Nai and Manoj Gupta
National University of Singapore

ACKNOWLEDGMENTS

We would like to take this opportunity to express our heartiest gratitude and thanks to all the people who have contributed to and assisted with the publication of this book. We would particularly want to express our sincere thanks to our families for their continual support and understanding, and to our co-workers, friends and students for their continual encouragement.

1

INTRODUCTION TO MAGNESIUM

This chapter introduces magnesium as an energy-efficient material that has the potential to replace steel, aluminum alloys, and some plastic-based materials. This is possible for a design engineer as the specific strength and stiffness of magnesium exceeds that of most commonly used metals and some plastic-based materials. Serving engineering applications since 1920s, magnesium was not the material of choice for many applications due to its high cost till about two decades back. Interest in magnesium-based materials is recently revived primarily because of its gradually reducing cost and the resolve of the scientists, researchers, and engineers to cut down energy consumption and greenhouse gas emissions.

1.1. INTRODUCTION

Over the years, with the increasing demand for economical use of scarce energy resources, skyrocketing crude oil prices (see Figure 1.1) [1], and ever-stricter control over emissions to lower environmental impact, industries are constantly searching for new, advanced materials as alternatives to "conventional" materials. The spike in crude oil

Magnesium, Magnesium Alloys, & Magnesium Composites, by Manoj Gupta and Nai Mui Ling, Sharon
© 2010 John Wiley & Sons, Inc.

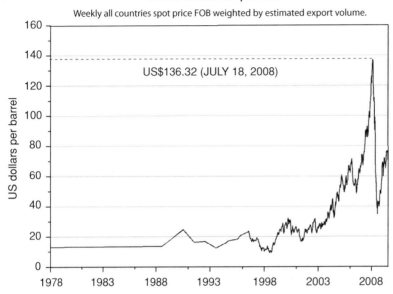

Figure 1.1. World crude oil prices. (Energy Information Administration, US)

price in July 2008 (see Figure 1.1) revealed the future trend of oil prices. Owing to this price rise, coupled with the depletion of energy resources with time, the choice of lightweight metals is the key and unavoidable solution for the future. Magnesium is one such promising lightweight metal, which is currently underutilized for engineering applications.

Magnesium is the sixth most abundant element in the earth's crust, representing 2.7% of the earth's crust [2]. Although magnesium is not found in its elemental form, magnesium compounds can be found worldwide. The most common compounds are magnesite ($MgCO_3$), dolomite ($MgCO_3 \cdot CaCO_3$), carnallite ($KCl \cdot MgCl_2 \cdot 6H_2O$), and also seawater [3]. Magnesium is the third most abundant dissolved mineral in the

TABLE 1.1. Density of commonly used structural materials [7, 8].

Materials	Density (g/cm^3)
Steel (cast iron)	7.2
Titanium	4.51
Aluminum	2.71
Magnesium	1.74
Structural plastic[a]	1.0–1.7

[a]The density value is dependent on the type and amount of reinforcements.

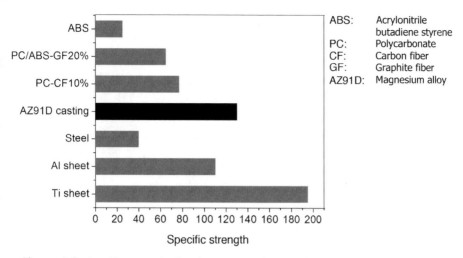

Figure 1.2. Specific strength of various structural materials. (Data extracted from [8].)

seawater (1.1 kg/m^3). Magnesium is the lightest of all structural metals. It has a density of 1.74 g/cm^3, which is approximately one-fourth the density of steel and two-thirds that of aluminum (see Table 1.1) [4–6]. Because of its low density and high specific mechanical properties (Figures 1.2 and 1.3), magnesium-based materials are actively pursued by companies for weight-critical applications.

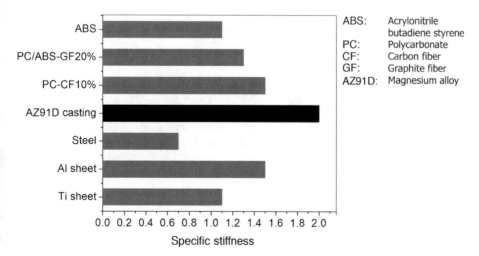

Figure 1.3. Specific stiffness of various structural materials. (Data extracted from [8].)

1.2. CHARACTERISTICS OF PURE MAGNESIUM [3]

1.2.1. Atomic Properties and Crystal Structure

Symbol	Mg
Element classification	Alkaline earth metal
Atomic number	12
Atomic weight	24.3050(6)
Atomic volume	14.0 cm^3/mol
Atomic radius	0.160 nm
Ionic radius	0.072 nm
Orbital electron states in free atoms	$1s^2, 2s^2, 2p^6, 3s^2$
Electrons per shell	2, 8, 2
Most common valence	2+
Crystal structure	Hexagonal close-packed (HCP)

1.2.2. Physical Properties

Density (at 20°C)	1.738 g/cm^3
Melting point	(650 ± 1)°C
Boiling point	1090°C
Linear coefficient of thermal expansion	
At 20–100°C	26.1×10^{-6} °C^{-1}
At 20–200°C	27.1×10^{-6} °C^{-1}
At 20–300°C	28.0×10^{-6} °C^{-1}
At 20–400°C	29.0×10^{-6} °C^{-1}
At 20–500°C	29.9×10^{-6} °C^{-1}
Thermal conductivity (at 27°C)	156 W m^{-1} K^{-1}
Specific heat capacity (at 20°C)	1.025 kJ kg^{-1} K^{-1}
Latent heat of fusion	360–377 kJ kg^{-1}
Latent heat of vaporization	5150–5400 kJ kg^{-1}
Latent heat of sublimation (at 25°C)	6113–6238 kJ kg^{-1}
Heat of combustion	24.9–25.2 MJ kg^{-1}
Coefficient of self-diffusion	
At 468°C	4.4×10^{-10} cm^2 s^{-1}
At 551°C	3.6×10^{-9} cm^2 s^{-1}
At 627°C	2.1×10^{-8} cm^2 s^{-1}

1.2.3. Electrical Properties

Electrical conductivity	38.6% IACS
Electrical resistivity (polycrystalline magnesium)	
At 20°C	44.5 nΩ m
At 316°C	92.8 nΩ m
At 593°C	139.5 nΩ m

TABLE 1.2. Mechanical properties of pure Mg at 20°C [3, 9–11].

Pure Magnesium	Annealed Sheet	Hand-Rolled Sheet	Sand Cast	Extruded	PM-Extruded	DMD-Extruded
0.2% Compressive yield strength (MPa)	69–83	105–115	21	34–55	92 ± 12^a	74 ± 4^b
0.2% Tensile yield strength (MPa)	90–105	115–140	21	69–105	132 ± 7^c	97 ± 2^d
Ultimate tensile strength (MPa)	160–195	180–220	90	165–205	193 ± 2^c	173 ± 1^d
Hardness HB^e	40–41	45–47	30	35	—	—

^{a}PM: powder metallurgy method, extruded at 350°C, extrusion ratio 20.25:1 [9].
^{b}DMD: disintegrated melt deposition method, extruded at 350°C, extrusion ratio 20.25:1 [11].
^{c}PM: powder metallurgy method, extruded at 250°C, extrusion ratio 20.25:1 [10].
^{d}DMD: disintegrated melt deposition method, extruded at 250°C, extrusion ratio 20.25:1 [10].
^{e}Using 10-mm diameter ball, 500-kg load.

1.2.4. Mechanical Properties

Table 1.2 shows the room-temperature mechanical properties of pure magnesium processed under different conditions.

1.3. APPLICATIONS

1.3.1. Automotive Applications

In the 1920s, magnesium parts made their way into racing cars. However, it was not until the 1930s that magnesium was used in commercial vehicles such as the Volkswagen (VW) Beetle. The VW Beetle, back then, contained more than 20 kg of magnesium alloy in the transmission housing and the crankcase.

Over the past decade, the increasing environmental and legislative pressures on the automotive industry to produce lighter, higher fuel efficiency, and higher performance vehicles have resulted in the surge in the use of magnesium. Widely used conventional steel parts are being replaced by new advanced materials such as magnesium, aluminum, and metal matrix composites. Leading automobile makers such as Audi, Volkswagen, DaimlerChrysler (Mercedes-Benz), Toyota, Ford, BMW, Jaguar, Fiat, Hyundai, and Kia Motors Corporation have used magnesium-based materials in their automotive parts. Figure 1.4 shows some of the actual magnesium automotive components. Figure 1.5 shows the NUS-FSAE Car using Mg alloy in wheel assembly. This car is built by a group of mechanical engineering students from the National University of Singapore (NUS),

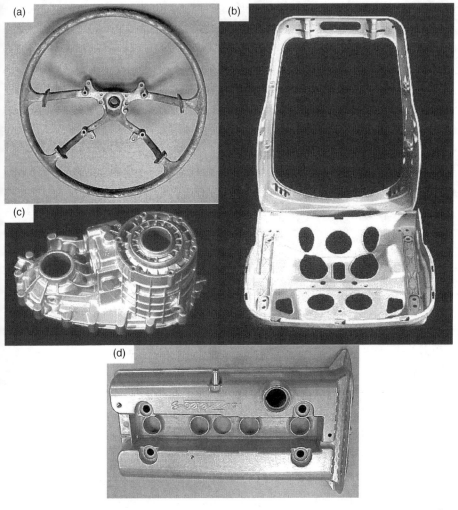

Figure 1.4. Magnesium automotive components: (a) magnesium steering wheel core for Toyota Camry weighing 0.75 kg, (b) seat support for Jaguar and Fiat models weighing 2.6 kg, (c) rear transfer case made from AZ91D weighing 2.7 kg, and (d) AZ91 magnesium alloy cam cover for Ford Zetec engine weighing 0.9 kg.

to participate in the FSAE (Formula Society of Automotive Engineers) competition in the United States.

In the VW Passat and Audi A4, magnesium parts are used in the gearbox housing [12]. In the Toyota Lexus, Carina, Celica, and Corolla, the steering wheels are made of magnesium [13]. In the Mercedes-Benz SLK, the fuel tank cover is made of a magnesium alloy. In Hyundai Azera (Grandeur) and Kia Amanti (Opirus), magnesium is also used in interior parts such as the seat frame, steering column housing, driver's air

Figure 1.5. NUS-FSAE Car using Mg alloy in wheel assembly. (Courtesy: Professor K. H. Seah, National University of Singapore.)

bag housing, steering wheel, and lock body [14]. Hyundai and Kia Motors Corporation project that the use of a magnesium seat frame translates to a 6 kg weight reduction per car (∼40% weight reduction by replacing steel with magnesium alloy). Thus, their annual consumption of magnesium was expected to increase from 670 tons in 2004 to 3700 tons in 2007 [14].

1.3.2. Aerospace Applications

In the aerospace industry, weight reduction is one of the most critical objectives due to the increasing need for emission reduction and fuel efficiency. The reduction in overall weight of the aircraft will result in fuel savings, which translates to savings in the total operational cost. Several weight reduction alternatives such as aluminum, fiber metal laminates, and low-density structural plastics have been introduced over the years. However, the limited advancement in the development of aluminum alloys has made further weight reduction a challenge. Fiber metal laminates are also high-cost materials and, hence, are only used for primary structures with the highest mechanical properties requirements. Moreover, low-density structural plastics have low impact and damage tolerance properties. They also exhibit inferior properties when subjected to temperature extremes. Thus, all these limitations have made magnesium an attractive alternative.

Magnesium-based materials have a long history of application in the aerospace industry. Over the years, magnesium-based materials are extensively used in both civil and military aircraft. Some applications include the thrust reverser (for Boeing 737, 747, 757, 767), gearbox (Rolls-Royce), engines, and helicopter transmission casings, etc. Military aircraft, such as the Eurofighter Typhoon, Tornado, and F16, also benefit from the lightweight characteristics of magnesium alloys for transmission casings.

There is also widespread use of magnesium in spacecraft and missiles due to the requirement for lightweight materials to reduce the lift-off weight. This is coupled with its high specific mechanical properties, ease of fabrication, and other attractive features such as its capability to withstand (i) elevated temperatures, (ii) exposure to ozone, and (iii) bombardment of high-energy particles and small meteorites. Large amount of magnesium (in the form of sheets) was used in the Titan, Agena, and Atlas intercontinental ballistic missiles [15].

1.3.3. Medical Applications

Magnesium alloys were first introduced as orthopedic biomaterials in the first half of the last century [16]. However, because of its low corrosion resistance, a large amount of hydrogen accumulates around the implant during the *in vivo* corrosion process, confining the widespread use of magnesium-based materials as biomaterials. Despite this, magnesium still possesses many attractive characteristics that make magnesium-based materials potential candidates to serve as implants for load-bearing applications in the medical industry.

Magnesium has a much lighter density than other implant materials (Table 1.3). It also has greater fracture toughness as compared to hydroxyapatite. Furthermore, as shown in Table 1.3, its elastic modulus and compressive yield strength values are more comparable to that of natural bone than the other commonly used metallic implants [17].

Magnesium is also present as a natural ion in the human body, whereby approximately 1 mol of magnesium is stored in a 70 kg adult human body and an estimated amount of half of the total physical magnesium is present in the bone tissue [17]. It also assists in many human metabolic reactions and is nontoxic to the human body. Magnesium

TABLE 1.3. Physical and mechanical properties of natural bone and some implant materials [17].

Materials	Density (g/cm^3)	Fracture Toughness (MPa m$^{1/2}$)	Elastic Modulus (GPa)	Compressive Yield Strength (MPa)
Natural bone	1.8–2.1	3–6	3–20	130–180
Ti alloy	4.4–4.5	55–115	110–117	758–1117
Co–Cr alloy	8.3–9.2	—	230	450–1000
Stainless steel	7.9–8.1	50–200	189–205	170–310
Magnesium	1.74–2.0	15–40	41–45	65–100
Hydroxyapatite	3.1	0.7	73–117	600

has good biocompatibility and it is biodegradable in human body fluid by corrosion, thus eliminating the need for another operation to remove the implant. All these desirable features make magnesium-based material a promising implant material [17–19].

In order to overcome the corrosion issues that limit the use of magnesium-based materials in orthopedics application, in recent years, much research efforts are focused to explore the use of different alloying elements in magnesium and surface treatments such as protective coatings on magnesium-based materials [17].

1.3.4. Sports Applications

In the sporting industry, it is important that the sports equipment matches up to the ever-increasing expectations of sports enthusiasts. The excellent specific strength and ability of magnesium alloys and magnesium composites to form intricate shapes resulted in many applications in sports-related equipment. For example, magnesium-based materials are used in the handles of archery bows, tennis rackets, and golf clubs (Figure 1.6).

Figure 1.6. Magnesium sports equipment: (a) golf club head is cast from high-quality magnesium (courtesy of www.thegolfdome.ca), (b) in-line skates with magnesium chassis (courtesy of www.skates.com), (c) tennis racquet with magnesium head (courtesy of www.courtsidesports.com), and (d) bicycle with magnesium frame (courtesy of www.segalbikes.eu).

Figure 1.7. Laptop with magnesium alloy
AZ91D casing.

The lightweight and excellent damping characteristics of magnesium-based materials have also made them a popular material choice in bicycle frames and the chassis of in-line skates (Figure 1.6). Bicycle frames made from magnesium alloys or composites are capable of absorbing shock and vibration [15], hence allowing the rider to exert less energy and enjoy a more comfortable ride.

1.3.5. Electronic Applications

The trend in the electronic equipment industry is to make products more personal and portable. Hence, it is important that the components that make up the equipment are lightweight and also durable. Magnesium-based materials meet the necessary requirements as they are as light as plastic, but exhibit great improvement in strength, heat transfer, and the ability to shield electromagnetic interference and radio frequency interference, as compared with their plastic counterparts [15]. Hence, as shown in Figure 1.7, magnesium-based materials are used in housings of cell phones, computers, laptops, and portable media players (such as the Apple iPod Nano magnesium case).

The ability to form magnesium alloys into complex shapes and the good heat dissipation and heat transfer characteristics of magnesium alloys also result in the use of magnesium alloys in heat sinks and the arms of the hard-drive reader [15]. Other examples of the use of magnesium include the housings of cameras (Figure 1.8) and digital image projection systems.

1.3.6. Other Applications

Optical Applications. Magnesium is commonly used to make the frame of eyewear because of its lightweight property. Other optical equipment that capitalizes on the

Figure 1.8. Magnesium housing of digital camera.

lightweight and optical stability attributes of magnesium includes the rifle scopes and binoculars.

Hand-Held Working Tools. In order to achieve higher working efficiency, it is desirable that the hand-held working tools are lightweight to allow greater portability. Hence, the low density of magnesium coupled with its resistance to impact and its ability to reduce noise and vibration makes it the material of choice for a wide range of hand-held working tools. Some examples include [15] the following:

(i) Magnesium chain saw housing
(ii) Magnesium housing and cylinder of pneumatic nail gun
(iii) Housings of gear and engine of hand-held tools
(iv) Handles of hand shears
(v) Housing of hand drills

1.4. SUMMARY

This chapter presents the potential of magnesium as an energy-efficient material. Its lightweight and high specific strength characteristics are favorable properties that have resulted in many applications in the automotive, aerospace, sports, and electronic industries. The future applications of magnesium-based materials are unlimited and depend on the vision and imagination of working engineers.

REFERENCES

1. US Energy Information Administration Website: http://tonto.eia.doe.gov/dnav/pet/hist/wtotworldw.htm (last accessed on December 20, 2009.).
2. H. Okamoto (1998) In A. A. Nayeb-Hashemi and J. B. Clark (eds) *Phase Diagrams of Binary Magnesium Alloys*. Metals Park, OH: ASM International.
3. M. M. Avedesian and H. Baker (ed.) (1999) *ASM Specialty Handbook—Magnesium and Magnesium Alloys*. Materials Park, OH: ASM International.

4. B. L. Mordike and K.U. Kainer (ed.) (1998) *Magnesium Alloys and Their Applications*. Frankfurt, Germany: Werkstoff-Informationsgesellschaft mbH.

5. B. L. Mordike and T. Ebert (2001) Magnesium: properties—applications—potential. *Materials Science Engineering A*, **302**, 37–45.

6. G. Neite, K. Kubota, K. Higashi, and F. Hehmann (2005) In R. W. Cahn, P. Haasen, and E. J. Kramer (eds) *Materials Science and Technology*, Vol. 8. Germany: Wiley-VCH.

7. W. D. Callister (2003) *Materials Science and Engineering: An introduction*. New York: Wiley.

8. J. F. King (2007) Magnesium: commodity or exotic? *Materials Science and Technology*, **23**(1), 1–14.

9. S. K. Thakur, M. Paramsothy, and M. Gupta (2010) Improving tensile and compressive strengths of magnesium by blending it with alumnium. *Materials Science and Technology*, **26**(1), 115–120.

10. S. F. Hassan and M. Gupta (2006) Effect of type of primary processing on the microstructure, CTE and mechanical properties of magnesium/alumina nanocomposites. *Composite Structures*, **72**, 19–26.

11. M. Paramsothy, M. Gupta, and N. Srikanth (2008) Improving compressive failure strain and work of fracture of magnesium by integrating it with millimeter length scale aluminum. *Journal of Composite Materials*, **42**(13), 1297–1307.

12. K. U. Kainer (ed.) (2003) *Magnesium—Alloys and Technologies*. Weinheim, Cambridge: Wiley-VCH.

13. D. Magers (1995) *Einsatzmöglichkeiten von Magnesium im Automobilbau. Leichtmetalle im Automobilbau (Sonderausgabe der ATZ und MTZ)*. Stuttgart: Franckh-Kosmos Verlags-GmbH.

14. J. J. Kim and D. S. Han (2008) Recent development and applications of magnesium alloys in the Hyundai and Kia Motors Corporation. *Materials Transactions*, **49**, 894–897.

15. H. E. Friedrich and B. L. Mordike (ed.) (2006) *Magnesium Technology—Metallurgy, Design Data, Applications*. Springer.

16. E. D. McBride (1938) Absorbable metal in bone surgery. *Journal of American Medical Association*, **111**(27), 2464–2467.

17. M. P. Staiger, A. M. Pietak, J. Huadmai, and G. Dias (2006) Magnesium and its alloys as orthopedic biomaterials: a review. *Biomaterials*, **27**, 1728–1734.

18. Y. W. Song, D. Y. Shan, and E. H. Han (2008) Electrodeposition of hydroxyapatite coating on magnesium alloy for biomaterial application. *Materials Letters*, **62**, 3276–3279.

19. Y. W. Song, D. Y. Shan, R. S. Chen, F. Zhang, and E. H. Han (2009) Biodegradable behaviors of AZ31 magnesium alloy in simulated body fluid. *Materials Science and Engineering C*, **29**(3), 1039–1045.

2

SYNTHESIS TECHNIQUES FOR MAGNESIUM-BASED MATERIALS

This chapter introduces the synthesis techniques for magnesium-based materials. These techniques are broadly classified under the liquid phase and solid phase processes. The advantages and disadvantages of synthesis techniques are presented and discussed. In essence, contents of this chapter illustrate that industries do not need to invest significantly into the infrastructural change in case they decide to switch to the synthesis of magnesium-based materials.

2.1. INTRODUCTION

A number of synthesis techniques have been attempted over the years to synthesize magnesium-based materials. The selection of the synthesis technique is very crucial as the resultant microstructural features are highly influenced by the processing parameters. For materials with the same constitution, with the adoption of different synthesis techniques, the synthesized materials will yield different properties. The synthesis techniques or processing methods can be classified into the following two categories [1–4]:

   (i) Liquid phase methods
   (ii) Solid phase methods

Magnesium, Magnesium Alloys, & Magnesium Composites, by Manoj Gupta and Nai Mui Ling, Sharon
© 2010 John Wiley & Sons, Inc.

Solid phase processed materials exhibit the following advantages when compared to the liquid phase processed counterparts:

(i) Refined microstructural features

(ii) Capability of producing metastable phases in the microstructure

(iii) Superior strength levels

However, their disadvantages include the following:

(i) High processing cost

(ii) Tedious handling of fine powders

(iii) Thickness limitations

(iv) Lower fracture toughness and ductility

The following sections briefly describe the different types of processes that are used to process/synthesize magnesium-based materials with special emphasis on composites.

2.2. LIQUID PHASE PROCESSES

Liquid phase processing techniques can be classified into [1–4] the following categories:

(i) Sand casting

(ii) Die casting

(iii) Squeeze casting

(iv) Semisolid metal (SSM) casting

(v) Stir casting

(vi) Spray forming

(vii) Melt infiltration method

(viii) *In situ* synthesis

2.2.1. Sand Casting

Sand casting is a very common casting method that is used to fabricate metallic parts. The part to be cast is produced by pouring molten metal into a sand mold cavity. Upon cooling of the mold, the metal solidifies and the cast part is removed from the mold. However, in order to produce satisfactory casting quality, it is important to take precaution to minimize metal–mold reactions as molten magnesium reacts readily with many molding materials [2]. This is achievable by

(i) minimizing moisture content of the sand and

(ii) the use of suitable inhibitors to the sand mixture that is used to make the mold and the cores.

Inhibitors such as sulfur, potassium fluoroborate, boric acid, and ammonium fluorosilicate can be used individually or in combination with other inhibitors. Several factors that affect the amount of inhibitor incorporated to the sand are [2]:

(i) the temperature of pouring the molten metal,

(ii) the type of alloy that is being cast, and

(iii) the thickness of the casting section.

With a higher pouring temperature, the metal–mold reaction is expected to increase, and hence, a higher amount of inhibitor needs to be added to the sand. Furthermore, when different alloys are used, the difference in the alloys' densities will also affect the amount of inhibitor to be added. Larger casting sections will result in a slower cooling rate, which results in a higher risk of "burn out" of the inhibitor. More volatile inhibitors are required to be refilled into the mold areas, as with use such inhibitors tend to deplete from the mold surface.

2.2.2. Die Casting

Die casting is also known as high-pressure die casting. During die casting, the molten metal is forced through a narrow gate to fill up the mold cavity at a very fast rate depending on the

(i) thickness of the wall,

(ii) type of alloy,

(iii) flow distance, and

(iv) temperature of the die.

During the solidification step, a high pressure ranging from 40 to 1000 MPa is applied onto the melt. This is to ensure that gas inclusions that are present in the melt are compressed and solidification shrinkage of the alloy is reduced. Moreover, as the cast metal solidifies at a fast cooling rate (100–1000°C/s), a fine-grained microstructure results [2]. In order to allow easy removal of the cast material from the mold, lubricants are applied onto the mold surface to reduce the sticking effect.

Die casting is widely used to produce thin-walled parts. It also has the following benefits:

(i) It has the capability of forming a near-net shape.

(ii) It facilitates a higher production rate.

(iii) With die casting, quite intricate shapes are possible.

(iv) Good strength can be realized from the part so made.

(v) It has the capability of providing good surface and dimensional precision.

(vi) Machining can be minimized.

Die casting can be routinely used to produce small- to medium-size parts. Magnesium and its alloys are commonly die cast using cold-chamber machines.

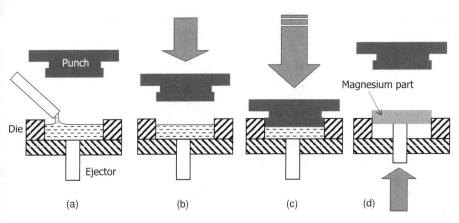

Figure 2.1. Schematic diagrams showing the process steps during liquid metal forging. (a) Pouring the molten magnesium metal into the mold cavity. (b) Lowering the punch. (c) Applying direct pressure. (d) Ejecting the solidified liquid forged magnesium part.

2.2.3. Squeeze Casting

Squeeze casting is a combination of the forging process and the casting process. There are two methods of squeeze casting: (i) direct squeeze casting and (ii) indirect squeeze casting [3, 4].

Direct squeeze casting is also called liquid metal forging (see Figure 2.1). The molten metal is poured into the lower half of a die and the upper punch is lowered to close up the die. The melt then solidifies under the application of high unidirectional pressure and the cast is formed. This casting method is widely used in the fabrication of metal matrix composites (MMCs), particularly fiber-reinforced MMCs. The processing variables governing the evolution of microstructures in squeeze cast MMCs include the following:

(i) Preheat temperature of fiber and melt

(ii) Infiltration speed and pressure

(iii) Spacing between fibers

When the temperature of the fiber is too low, castings of inferior quality (porous and poorly infiltrated castings) result. However, too high temperatures also result in excessive metal–fiber reaction, which degrades the properties of the cast. Furthermore, it is essential to have a threshold pressure to initiate the liquid metal flow through a fibrous preform or powder bed. This is to overcome the viscous friction of molten metal moving through the reinforcements. In the case of discontinuously reinforced MMCs, reinforcements such as particulates or whiskers are mixed with the molten metal before the squeeze casting process [5].

For direct squeeze casting, although the absence of the runner system results in high material yield, it also traps the impurities that are initially present in the melt, and hence, in the resultant cast part. The application of high pressure during the solidification step assists in virtually eliminating solidification shrinkage, gas-associated porosities, and metal-reinforcement interfacial voids.

Indirect squeeze casting is performed in a way similar to die casting, whereby the molten metal is poured into the sleeve of squeeze casting equipment. By controlling the speed of the plunger, the speed of the molten metal filling is manipulated. Effort must be made to carefully select the filling speed so as to minimize any turbulent flow, to achieve pore-free castings. This casting method is also suitable for the infiltration of pre-forms for the synthesis of MMCs. However, in comparison with direct squeeze casting, there is more material loss in indirect squeeze casting, thus reducing the material yield.

2.2.4. SSM Casting

Of all the applications of Mg alloys, the majority are fabricated using the high-pressure die casting method. This is due to its high production efficiency, high production volume, and low production cost. Despite all these benefits, the parts fabricated using the die casting method still possess defects. These parts are found to have porosity resulting from gas entrapment during the die filling step, and the parts also experience hot tearing during the solidification in the die [6–8]. These defects will degrade the material's mechanical performance and hinder further property enhancement by the successive heat treatment process.

One way to overcome the above-mentioned issues with high-pressure die casting of magnesium-based materials is to use the SSM casting processes. SSM casting is also known as thixomolding, rheocasting, thixocasting, or thixoforming. Some of the advantages of the SSM processes include the following [9, 10]:

(i) Components with low porosity level
(ii) Components of consistent mechanical properties
(iii) Ability to produce components of complex shapes
(iv) Longer die life

The heat content of a semisolid slurry is lower and has a higher viscosity as compared with that of the same alloy in liquid state. The lower heat content allows faster cooling and greater thermal efficiency, and also aids in prolonging the die life. Moreover, with higher viscosity, the slurry can fill the die cavity with less turbulence and gas entrapment, leading to resultant components of superior properties [11]. Some drawbacks of the SSM process include a more expensive feedstock and the immediate recyclability issue of the scrap.

2.2.4.1. Thixomolding. Thixomolding is an SSM process that is specially developed for magnesium alloys [11]. It is similar to the plastic injection molding process and is a contender to compete with the hot chamber die casting for the production of thin-walled parts. This method involves the use of magnesium alloy chips as raw material. The chips are fed into a heated screw with the help of a volumetric feeder and are slowly heated up to reach a temperature below the liquidus temperature. The screw drives the chips down the barrel as they are heated to the semisolid temperature range. In order to minimize oxidation of the chips, the process is carried out under inert argon atmosphere. Heating coupled with the shearing forces provided by the rotation of the screw generate the semisolid slurry. The resulting slurry is subsequently injected by the forward movement of the screw into the die.

2.2.4.2. Rheocasting. Rheocasting is the casting of semisolid slurries, which is conducted within the liquidus–solidus range of the metallic alloy. In contrast to the thixomolding process, this process uses molten alloy as the starting material, thus eliminating the demand of specially prepared feedstock materials [4, 12].

For synthesizing composite materials [12], the addition of reinforcement is carried out within the semisolid temperature range. Following the incorporation of the reinforcements, stirring with the use of mechanical stirrer is carried out and the homogenized slurry is poured into a mold. The slurry characteristic of the matrix during stirring allows the addition of reinforcement during solidification. The ceramic particulates are mechanically entrapped initially and are prevented from agglomeration by the presence of primary alloy solid particles. Subsequently, the reinforcement particles interact with the liquid matrix to allow bonding. Furthermore, continuous deformation and breakdown of solid phases during agitation prevent particulate agglomeration and settling [12].

In comparison with conventionally cast parts, rheocast components possess the following advantages [13]:

(i) Homogeneous distribution of microporosity (if any)

(ii) Less tendency of shrinkage pipe and crack formation

(iii) Less tendency of micro- and macrosegragation

(iv) A fine and nondendritic grain structure

2.2.5. Stir Casting

Stir casting can be used to synthesize both magnesium alloys and their composites. In the case of composites, for example, the reinforcing phase is incorporated into a molten metallic matrix using various proprietary techniques. This is followed by mixing and eventual casting of the resulting composite mixture into either shaped components or billets [1, 12]. Figure 2.2 shows a schematic diagram of the stir casting process.

Besides reaching the advanced stage of development, various difficulties are encountered during stir casting of MMCs, and these include

(i) agglomeration of the reinforcing phase,

(ii) settling or floating of reinforcing phase (depending on the density of reinforcing phase and that of metallic matrix),

(iii) fracture of reinforcing phase during agitation, and

(iv) extensive interfacial reactions between the reinforcing phase and the metallic matrix.

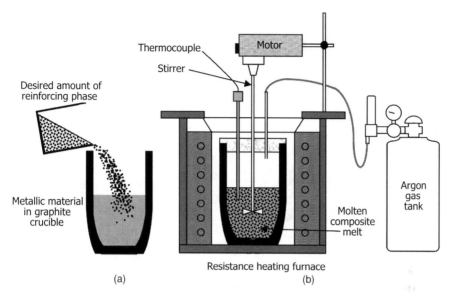

Figure 2.2. A schematic diagram showing (a) the incorporation of reinforcing phase into metallic material prior to heating and (b) the stir casting setup.

The initial difficulties indicated above have been extensively researched on and addressed successfully over the last 30 years.

2.2.6. Spray Forming

Spray forming is also known as spray casting or spray atomization and deposition in the industrial and academic field [1, 3, 14]. It is used to cast near-net shape metal parts. The spray forming process consists of two key stages (see Figure 2.3):

   (i) Atomization of the melt
   (ii) Spray deposition

In spray forming, the metal (monolithic or alloy or composite) is melted in an induction furnace or resistance furnace. For the synthesis of composite material, reinforcing particles can also be introduced outside the crucible and/or into molten spray. In order to achieve uniform distribution of the reinforcements in the molten metal, the melt is thoroughly mixed prior to pouring. Nonhomogeneous distribution of the reinforcements in the matrix will greatly compromise the properties of the resultant composite [1]. When resistance heating is used, an agitator (e.g., a mechanical stirrer) is inserted into the molten composite melt to facilitate uniform mixing. When RF (radio-frequency) induction heating is used, the presence of the strong electromagnetic interactions in the molten composite melt provides the adequate agitation required.

    When the melt is sufficiently mixed, it is subsequently released as a thin free-falling stream. This molten metallic stream is then impacted and disintegrated into droplets by

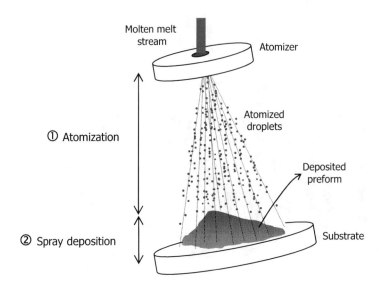

Figure 2.3. A schematic diagram showing the two key stages of spray forming.

highly energetic inert gas or water jets. These droplets are then accelerated and propelled away from the region of atomization by the atomizing gas/water. As schematically shown in Figure 2.3, the paths of the droplets are interrupted by the substrate that collects and solidifies the droplets into a preform. When the droplets strike the substrate, most of them are in two-phase (solid–liquid) form and with continuous deposition a preform is gradually built up [14].

2.2.7. Melt Infiltration Method

The melt infiltration method involves holding a porous body of the reinforcing phase in a mold and infiltrating it with the molten metallic material, which flows through the interstices to fill the pores, thus producing a composite material. This method can be divided into two categories [1, 3]:

  (i) Pressure-assisted infiltration
  (ii) Pressure-less infiltration

For pressure-assisted infiltration, either an inert gas or a mechanical device is used as the pressurizing medium. The composite produced using this method generally features a near pore-free matrix. There are also some drawbacks associated with this method:

  (i) Reinforcement preform damage or breakage during infiltration
  (ii) Microstructural heterogeneity

In view of these disadvantages, other types of forces such as ultrasonic vibration, electromagnetic force, and centrifugal force are used to more effectively force the molten metal into nonwetting reinforcement preforms [15].

In pressure-less infiltration, the liquid metal infiltrates a porous reinforcement pre-form without an external pressure or vacuum. This process is also known as spontaneous infiltration. In comparison with the pressure-assisted infiltration method, the composites formed exhibit a higher level of porosity.

When using fiber preforms of low volume fraction (less than 10%), it is challenging to effectively infiltrate molten metal into the fiber preform as the preform may be excessively deformed [16]. Furthermore, the melt infiltration method does not allow rapid cooling of the molten metal upon infiltration. Hence, undesirable interfacial reactions between the reinforcement and metallic matrix may take place. This can result in the degradation of the fiber and poor strength of the resultant composite.

2.2.8. *In Situ* Synthesis

The *in situ* reaction method refers to the process in which the reaction between the raw materials is utilized to synthesize the reinforcement in the metal matrix [3]. This can be achieved through (i) gas–liquid, (ii) liquid—liquid, and (iii) solid–liquid reactions on the basis of phase diagram principles. In order to obtain the desirable end product (*in situ* composites), it is important to have a good understanding of thermodynamics and reaction kinetics. Using the *in situ* reaction method, the reinforcement produced tends to be fine, well distributed, and exhibit good interfacial bonding and cleaner interface.

In a study by Matin et al. [17], for example, magnesium reinforced by *in situ* particulates was successfully synthesized using the reaction between pure magnesium, potassium boron tetrafluoride (KBF_4), and potassium titanium hexafluoride (K_2TiF_6) compounds. Through the chemical reaction between the compounds and magnesium, titanium borides reinforcement particles were formed.

2.3. SOLID PHASE PROCESS

The solid phase process involves the synthesis of materials at a temperature below the solidus temperature of the matrix phase. The powder metallurgy (PM) process is a solid phase process [1, 12]. It is divided into four key stages: (i) mixing of powder, (ii) powder consolidation, (iii) sintering, and (iv) secondary processing. Figure 2.4 presents the processing flowchart for metal-based composites for the PM route.

This flowchart is also applicable to the synthesis of metallic alloys. For synthesis of metallic alloys, preweighed powders of base metal and alloying elements are taken for the blending or mechanical alloying (MA) step. The rest of the steps remain similar.

2.3.1. Blending

Blending is the mixing step of the PM route. It normally involves dry mixing the metallic loose powder particles with the desired amount of reinforcing phase in the case of synthesis of composite material. Figure 2.5 shows the photograph of typical blending equipment.

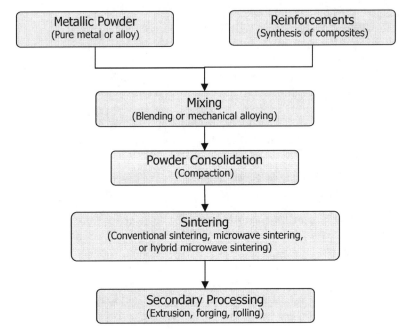

Figure 2.4. Manufacturing process flowchart for metal-based materials using the powder metallurgy route.

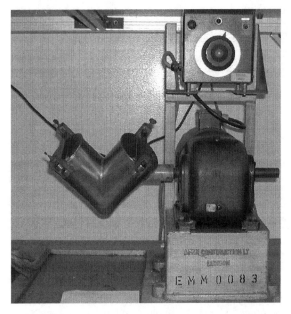

Figure 2.5. Photo showing V-blending equipment.

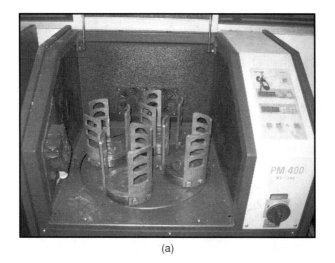

(a)

(b)

<u>Figure 2.6.</u> Photos showing (a) a typical planetary ball mill and (b) the required accessories.

2.3.2. Mechanical Alloying

MA is a solid-state powder processing technique. It is a ball-milling process that involves repeated welding and fracturing of the powder mixture in the ball mill. The powder mixture undergoes high-energy collision from the balls [18]. Figure 2.6 shows the photographs of a typical planetary ball mill and the accessories needed for the ball-milling process.

Prior to MA, the powder mixture (which consists of elemental powder particles, alloying powder particles, reinforcing powder particles, and process control agent) is placed in a container (which is also known as the jar, bowl, or vial) together with some steel or ceramic balls. A process control agent is used to avoid excessive welding in

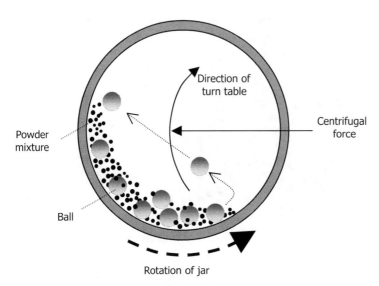

Powder mixture

Ball

Direction of turn table

Centrifugal force

Rotation of jar

Figure 2.7. Schematic diagram showing the motion of the powder mixture and balls inside the jar during the ball-milling process.

the metallic powder, which leads to the formation of lumps during the ball-milling step. Often stearic acid in the range of 1–3% is used. When processing, the jars are rotated (in clockwise and anticlockwise directions) and the balls collide with the powder mixture resulting in high-energy impact. The ball mill is normally made up of a turntable and two or four jars. During processing, the turntable and the jars rotate about their respective axis and in opposing directions. Thus, centrifugal forces are created. Figure 2.7 shows the schematic view of the motions of the balls and powder particles in a jar during the ball-milling process.

2.3.3. Powder Consolidation (Compaction)

Compaction is the powder-pressing step of the PM route. Commonly used powder-pressing processes can be classified into [19] two types:

(i) Uniaxial pressing
(ii) Isostatic pressing

For uniaxial pressing, firstly the loose powder is poured into a metal die cavity. With the applied pressure directed in a single direction, the powder is compacted into a solid piece. The compacted piece is subsequently removed from the die as shown in Figure 2.8.

For uniaxial pressing, there exists the cold compaction and hot compaction approaches. The setup for hot compaction consists of the compacting die, punch, heating

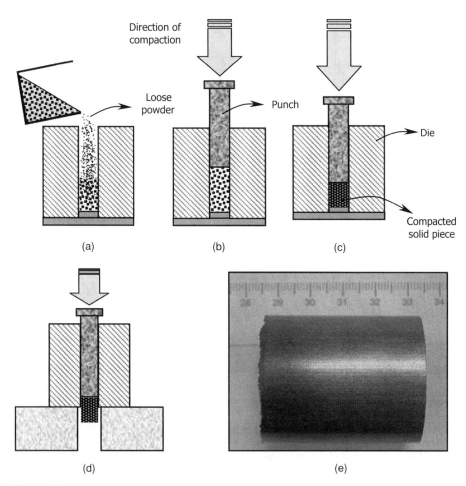

Figure 2.8. Schematic diagrams showing the uniaxial powder-pressing steps. (a) Filling the die cavity with loose powder particles. (b), (c) Applying uniaxial pressure to compact the powder. (d) Removing the compacted solid piece from the die. (e) Compacted magnesium nanocomposite billet.

element, and an oxidation protection device [18]. The punch is often installed with a cooling element to minimize softening.

Figure 2.9 shows the schematic representation of the hot compaction setup. When mechanically alloyed powder is to be compacted, it is essential to use inert gases to prevent oxidation due to the presence of very fine particles and the long compaction duration over elevated temperature.

For isostatic powder pressing, there is a pressure barrier between the pressurizing medium and the powder, whereby the powder is confined within a flexible membrane or hermetic container. This is unlike the case of uniaxial pressing where the loose powder is in direct contact with the die. In isostatic pressing, the pressure is applied from all

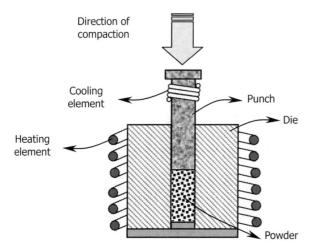

Figure 2.9. Schematic diagram showing the hot compaction setup.

directions—thus the term "isostatic"—and the pressurizing medium used is either a gas or liquid. Uniform compaction pressure is achieved throughout the powder compact, resulting in a uniform density distribution in the end product.

There are two types of isostatic powder-pressing methods:

(i) Cold (or room temperature) isostatic pressing (CIP)
(ii) Hot isostatic pressing (HIP)

In uniaxial pressing using the cold die compaction method, it is almost impossible to achieve uniform densification because of the frictional force between the particles and the die [18]. This can be minimized using the cold isostatic pressing method. However, the isostatic pressing method, due to its higher production cost and slower processing speed, is only used to produce small quantities and simple-shaped materials. Figure 2.10 shows the photo of a typical uniaxial powder compaction equipment.

2.3.4. Sintering Methods

Sintering is one of the key processing steps of the PM route [1, 12, 18]. It involves heating the compacted powder below the material's melting point until the powder particles coalesce. During the initial sintering stage, necks are formed along the particle contacts. As the sintering process progresses, the voids that are formed internally begin to diminish in size as the compacted solid becomes denser (see Figure 2.11).

There are several heating modes to sinter materials, which can be broadly grouped into [20] the following three categories:

(i) Conventional resistance heating
(ii) Pure microwave heating
(iii) Hybrid microwave heating

Figure 2.10. Photograph of uniaxial powder compaction equipment.

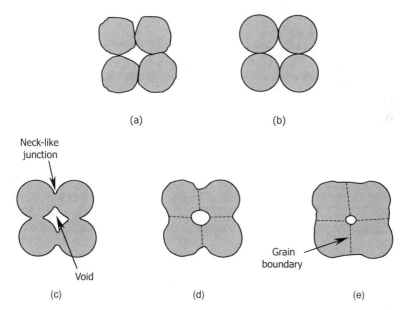

Figure 2.11. Schematic diagrams showing the sintering stages. (a) Compacted powder particles before subjected to sintering. (b), (c) Particles adhere to each other and the formation of void as sintering begins. (d), (e) As sintering progresses, the voids change in shape and diminish in size.

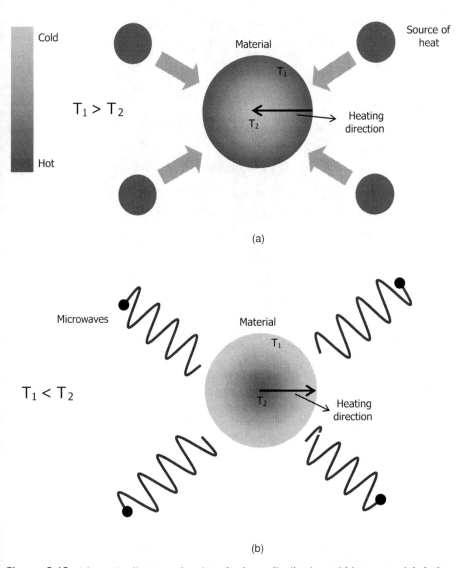

Figure 2.12. Schematic diagrams showing the heat distribution within a material during (a) conventional resistance heating, (b) pure microwave heating, and (c) hybrid microwave heating with the use of microwave susceptors.

Figure 2.12 shows the schematic representations of the heat distribution within a material during the different heating modes.

2.3.4.1. Conventional Sintering. In conventional heating, thermal energy is transferred from the outer surface of the material to the inner surface through conduction, convection, and radiation of heat that is generated by external sources such as a

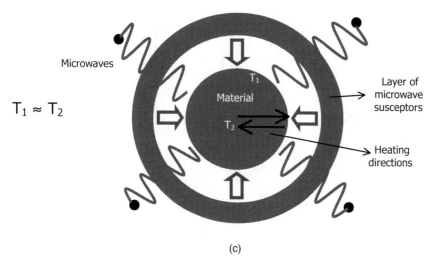

(c)

Figure 2.12. *(Continued)*

resistive heating element. As the penetration depth of infrared radiation is very small in most solids, the energy deposition is limited to the outer surfaces of the material. This results in lower temperature in the core of the material as compared with that at the material's surface during conventional heating (see Figure 2.12a). Accordingly, relatively poor microstructural characteristics of the core can be expected. Figure 2.13 shows the photograph of a conventional resistance sintering tube furnace.

2.3.4.2. Microwave Sintering. In pure microwave heating, heat is generated from within the material and radiates outward because of the penetrative power of microwaves. The microwave energy is absorbed by the material and is not dependent on the heat transfer from the outer surfaces. This results in higher temperature in the core of the material as compared with that at the material's surface (as outer surface loses heat to the surroundings) during pure microwave heating (see Figure 2.12b). Accordingly, relatively poor microstructural characteristics of the surface can be expected.

In order to circumvent the disadvantage of heating using either conventional heating or microwaves only (see Figure 2.12c), the hybrid microwave heating method was developed and used [20, 21]. In hybrid microwave heating, microwave susceptors such as SiC particles/rods are used to assist in the reduction of the thermal gradient during the sintering of materials.

For hybrid microwave sintering, Gupta et al. [21] used a 900-W, 2.55-GHz SHARP microwave oven with SiC as the microwave susceptor material [6, 11, 12] to successfully sinter a wide range of unreinforced and reinforced metallic materials. Figure 2.14a shows the hybrid heating experimental setup used. A layer of SiC powder was placed in between the inner and outer crucibles as shown in Figure 2.14b.

Figure 2.13. Photograph of the tube furnace (Carbolite CTF)—conventional resistance sintering equipment.

The preforms were nonisothermally sintered. Temperature calibration of the sintering setup was performed beforehand using a thermocouple in order to determine the appropriate sintering duration for the material. The set time included only heating time and there was no holding time to minimize or eliminate bulk and/or localized melting of material. SiC powder (contained within a microwave-transparent ceramic crucible) absorbs microwave energy readily at room temperature and is heated up quickly, providing the radiant heat to heat the preforms externally, while the compacted preforms absorb microwaves and are heated from within (see Figure 2.12c).

During hybrid microwave sintering, as the heating is in two directions, microstructural nonuniformity is minimized. This also ensures enhanced properties and reduced sintering time. Thus, significant energy savings can be achieved using microwave heating as microwave energy is transferred directly to the metallic material, in contrast to conventional heating whereby energy is expended in heating the resistive coils, surrounding atmosphere, and the inner walls of the tube furnace. The use of external susceptors (such as SiC susceptors here) during hybrid microwave sintering also assists in achieving a fast heating rate and provides radiant heat to the metallic material externally, thus reducing the thermal variation in the compacted preforms. Figure 2.15 shows the photographs of a sintered magnesium powder billet and the extruded rod after secondary processing.

Table 2.1 shows a simple analysis of energy consumption to sinter magnesium- and aluminum-based materials using a microwave oven (with a hybrid microwave heating setup) and a conventional tube furnace. Using the hybrid microwave sintering approach, Wong et al. [22] reported a significant energy savings of up to 96% when compared

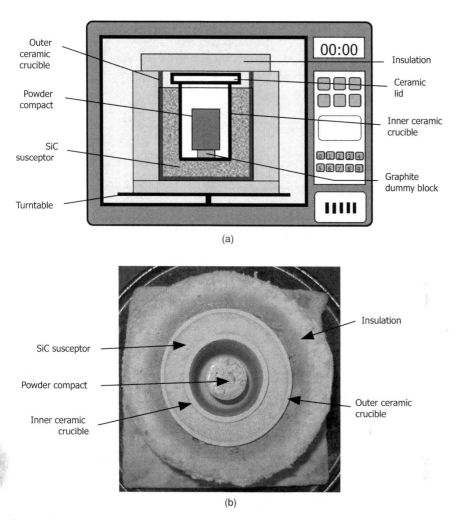

Figure 2.14. The hybrid microwave heating experimental setup used. (a) Schematic diagram and (b) photograph showing the top view of the setup (without the lid and top insulation).

with the conventional sintering approach (using the Carbolite tube furnace). This will definitely be economically viable for industries and environment friendly in terms of lower CO_2 emission.

Another key difference between the various sintering approaches is the sintering atmosphere. Conventional sintering is carried out in an inert argon atmosphere, whereas hybrid microwave sintering can be conducted in air without any inert protective atmosphere even in the case of highly reactive material such as magnesium. It is also interesting to note that despite the absence of an inert atmosphere during hybrid microwave sintering, the end properties of the hybrid microwave sintered materials are not compromised [21], and this indicates further cost savings.

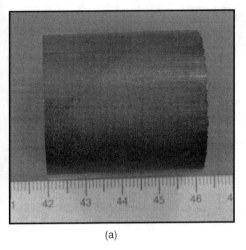
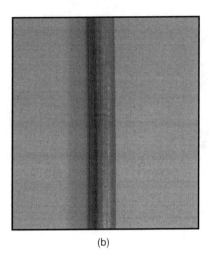

(a)   (b)

Figure 2.15. Photographs of (a) the sintered magnesium powder billet and (b) the 7-mm-diameter rod after it is subjected to the secondary extrusion process.

2.4. DISINTEGRATED MELT DEPOSITION METHOD

The disintegrated melt deposition (DMD) method combines the advantages of conventional casting and the spray process. Unlike spray processes, the DMD technique employs higher superheat temperatures and lower impinging gas jet velocity with the end product being only bulk alloy/composite material. It was developed in the National University of Singapore (NUS) in 1994 and is one of the most cost-effective processes tested successfully to synthesize aluminum- and magnesium-based monolithic, reinforced, and functionally gradient materials [23–29].

For the synthesis of Mg-based materials, the Mg material and the desired amount of reinforcements are firstly weighed and placed in a graphite crucible. The crucible was equipped with an arrangement for bottom pouring. Figure 2.16a shows the schematic

TABLE 2.1. Simple analysis of energy consumption to sinter magnesium- and aluminum-based materials [22].

Heating Source	Power (kW)	Sintering Duration (min)	Consumption of Energy (kW h)	Energy Savings (%)
Sharp microwave with hybrid sintering setup	1.6[a]	25	0.7	96[b]
Carbolite tube furnace	6	170	17	—

[a]Based on the AC power needed for the operation of magnetron.
[b]In comparison with the tube furnace, assuming the power stays ON during the heating and holding period.

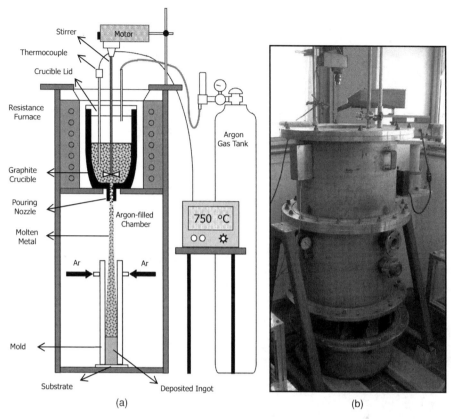

Figure 2.16. (a) Schematic diagram of the DMD setup. (b) Photograph of the actual DMD setup developed at the National University of Singapore.

diagram of the DMD experimental setup used, and Figure 2.16b shows the photograph of the actual DMD setup developed in NUS.

The materials are normally superheated to 750°C (1023 K) in a resistance furnace under an inert argon environment. Upon reaching the superheat temperature, the molten composite melt is mechanically stirred using a twin-blade mild steel impeller (with pitch 45°) to facilitate the incorporation and uniform distribution of reinforcement materials in the metallic matrix. The impeller is coated with a water-based zircon wash (Zirtex 25: 86% ZrO_2, 8.8% Y_2O_3, 3.6% SiO_2, 1.2% K_2O and Na_2O, and 0.3% trace inorganic) to prevent any possible iron contamination of the melt. The melt is then released through a 10-mm-diameter orifice at the base of the crucible (see Figure 2.16a). The composite melt is subsequently disintegrated by two jets of argon gas orientated normal to the melt stream. The argon gas flow rate is maintained at 25 L/min. The disintegrated composite melt is subsequently deposited onto a metallic substrate, which is situated at the base. Figure 2.17 shows the photographs of the magnesium-based material in its as-cast form and after secondary processing.

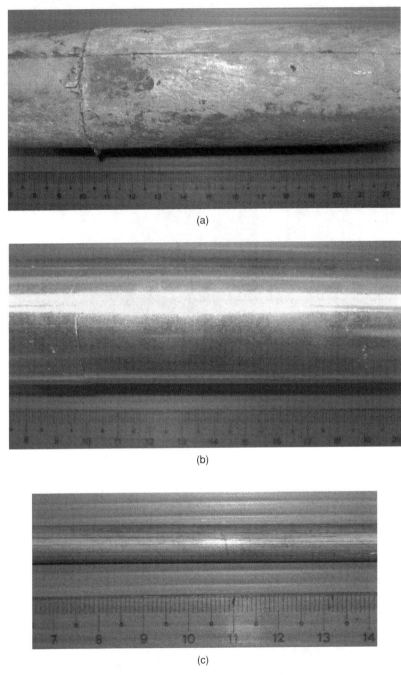

Figure 2.17. Photographs of (a) the as-cast magnesium-based material using the DMD technique, (b) the magnesium-based material after machining, revealing its surface, which is free from macropores, and (c) the magnesium-based rod after it is subjected to hot extrusion.

For the synthesis of functionally gradient materials, the DMD method can also be adopted. Nai et al. [23, 24] demonstrated that with the use of different stirrer geometries (twin bladed, four bladed, and circular shaped) [24] and the judicious control of stirring speed [23], aluminum–silicon carbide based functionally gradient materials are successfully fabricated using the DMD method. The same approach can be employed to synthesize Mg-based functionally gradient materials.

2.5. MECHANICAL DISINTEGRATION AND DEPOSITION METHOD

The mechanical disintegration and deposition (MDD) method is developed to further minimize the cost of the spray processes while retaining the scientific advantages

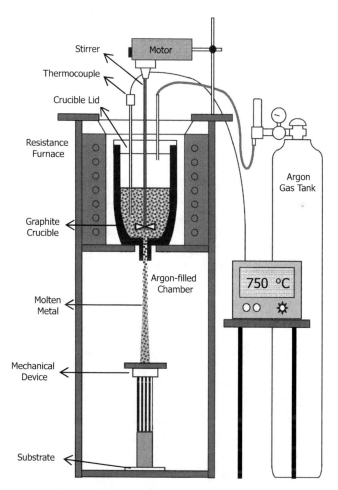

Figure 2.18. A schematic diagram of the mechanical disintegration and deposition setup.

associated with the spray methods. Unlike the conventional disintegration techniques that use inert gases for disintegration, this method employs a mechanical device to disintegrate the molten composite slurry before the deposition step, resulting in the synthesis of bulk composite material. It has been used to successfully synthesize unreinforced and reinforced magnesium-based materials [30].

For the synthesis of Mg-based materials, firstly the Mg material and the desired amount of reinforcements are carefully weighed and placed in a graphite crucible with an arrangement for bottom deposition inside an electric resistance furnace (see Figure 2.18).

The materials are heated to a superheat temperature of 750°C (1023 K) under an inert argon atmosphere. Upon reaching the desired temperature, a twin-blade mild steel impeller (with pitch 45°) is used to mechanically mix the molten mixture to ensure uniform distribution of the reinforcements in the Mg matrix prior to pouring. The impeller is coated with Zirtex 25 to prevent any possible iron contamination of the melt. After mechanical mixing, the molten melt is released through a 10-mm diameter orifice at the base of the graphite crucible. With the help of a mechanical device, which is located below the base of the crucible, the molten composite melt is then disintegrated into multiple uniform streams. This mechanical device is preheated at 1000°C (1273 K) for an hour prior to usage. The disintegrated melt is subsequently deposited onto a metallic substrate to obtain an ingot of desired dimension.

2.6. SUMMARY

This chapter illustrates that magnesium-based materials can be synthesized using a variety of techniques that are conventionally used for other metal-based materials. Magnesium-based materials do not require special and expensive setups for synthesis. However, as magnesium-based materials are comparatively more reactive, it is essential that preventive mechanisms such as the use of fluxes and inert atmospheric conditions be used to avoid their interactions with oxygen-containing environment.

REFERENCES

1. D. J. Lloyd (1994) Particle reinforced aluminum and magnesium matrix composites. *International Materials Reviews*, **39**(1), 1–23.

2. M. M. Avedesian and H. Baker (ed.) (1999) *ASM Specialty Handbook—Magnesium and Magnesium Alloys*. Materials Park, OH: ASM International.

3. N. Chawla and K. K. Chawla (2006) *Metal Matrix Composites*. New York: Springer.

4. H. E. Friedrich and B. L. Mordike (ed.) (2006) *Magnesium Technology—Metallurgy, Design Data, Applications*. Berlin: Springer.

5. P. K. Rohatgi (1993) Metal-matrix composites. *Defence Science Journal*, **43**(4), 323–349.

6. A. Balasundaram and A. M. Gokhale (2001) In J. N. Hryn (ed.) *Magnesium Technology*. Warrendale, PA,: TMS, pp. 155–159.

7. H. Kaufmann and P. J. Uggowitzer (2001) Fundamentals of the new rheocasting process for magnesium alloys. *Advanced Engineering Materials*, **3**(12), 963–967.

8. Z. Fan (2005) Development of the rheo-diecasting process for magnesium alloys. *Materials Science and Engineering A*, **413–414**, 72–78.

9. M. C. Fleming (1991) Behavior of metal alloys in the semisolid state. *Metallurgical Transactions A*, **22**, 957–981.

10. D. H. Kirkwood (1994) Semisolid metal processing. *International Materials Review*, **39**(5), 173–189.

11. D. H. Kirkwood (2008) In K. H. J. Buschow, R. W. Cahn, M. C. Flemings, B. Ilschner, E. J. Kramer, S. Mahajan, and P. Veyssiere (eds), *Encyclopedia of Materials: Science and Technology*. New York: Elsevier, pp. 8432–8437.

12. I. A. Ibrahim, F. A. Mohamed, and E. J. Lavernia (1991) Particulate reinforced metal matrix composites—a review. *Journal of Materials Science*, **26**(5), 1137–1156.

13. A. Tissier, D. Apelian, and G. Regazzoni (1990) Magnesium rheocasting: a study of processing-microstructure interactions. *Journal of Materials Science*, **25**(2), 1184–1196.

14. E. J. Lavernia and Y. Wu (1996) *Spray Atomization and Deposition*. New York: John Wiley & Sons, Ltd.

15. R. Asthana (1997) Cast metal-matrix composites I: Fabrication techniques. *Journal of Materials Synthesis and Processing*, **5**(4), 251–278.

16. R. K. Everett and R. J. Arsenault (eds) (1991) *Metal Matrix Composites: Processing and Interface*. New York: Academic Press, pp. 1–16.

17. M. A. Matin, L. Lu, and M. Gupta (2001) Investigation of the reactions between boron and titanium compounds with magnesium. *Scripta Materialia*, **45**(4), 479–486.

18. L. Lu and M. O. Lai (1998) *Mechanical Alloying*. Boston: Kluwer.

19. W. D. Callister (1997) *Materials Science and Engineering: An Introduction*. New York: John Wiley & Sons, Ltd., 427 pp.

20. M. Gupta and W. L. E. Wong (2007) *Microwaves and Metals*. Singapore: John Wiley & Sons, Ltd.

21. M. Gupta and W. L. E. Wong (2005) Enhancing overall mechanical performance of metallic materials using two-directional microwave assisted rapid sintering. *Scripta Materialia*, **52**(6), 479–483.

22. W. L. E. Wong and M. Gupta (2007) Improving overall mechanical performance of magnesium using nano-alumina reinforcement and energy efficient microwave assisted processing route. *Advanced Engineering Materials*, **9**(10), 902–909.

23. S. M. L. Nai and M. Gupta (2002) Influence of stirring speed on the synthesis of Al/SiC based functionally gradient materials. *Composite Structures*, **57**, 227–233.

24. S. M. L. Nai and M. Gupta (2003) Synthesis and characterization of free standing, bulk Al/SiC$_p$ functionally gradient materials: Effects of different stirrer geometries. *Materials Research Bulletin*, **38**(11–12), 1573–1589.

25. Q. B. Nguyen and M. Gupta (2008) Increasing significantly the failure strain and work of fracture of solidification processed AZ31B using nano-Al$_2$O$_3$ particulates. *Journal of Alloys and Compounds*, **459**(1–2), 244–250.

26. C. S. Goh, J. Wei, L. C. Lee, and M. Gupta (2006) Simultaneous enhancement in strength and ductility by reinforcing magnesium with carbon nanotubes. *Materials Science and Engineering A*, **423**(1–2), 153–156.

27. S. F. Hassan and M. Gupta (2007) Effect of nano-ZrO$_2$ particulates reinforcement on microstructure and mechanical behavior of solidification processed elemental Mg. *Journal of Composite Materials*, **41**(21), 2533–2543.

28. W. L. E. Wong and M. Gupta (2005) Enhancing thermal stability, modulus and ductility of magnesium using molybdenum as reinforcement. *Advanced Engineering Materials*, **7**(4), 250–256.

29. M. Paramsothy, N. Srikanth, and M. Gupta (2008) Solidification processed Mg/Al bimetal macrocomposite: microstructure and mechanical properties. *Journal of Alloys and Compounds*, **461**(1–2), 200–208.

30. M. K. K. Oo, P. S. Ling, and M. Gupta (2000) Characteristics of Mg-based composites synthesized using a novel mechanical disintegration and deposition technique. *Metallurgical and Materials Transactions A*, **31**(7), 1873–1881.

3

MAGNESIUM ALLOYS

This chapter attempts to introduce the fundamentals related to magnesium alloys. The influence of various elements in altering the characteristics of magnesium is indicated. Alloy and temper designations as relevant to magnesium-based alloys are highlighted. Chemical compositions, characteristics, and properties of cast and wrought alloys are given in a simplistic fashion. Magnesium alloys for elevated temperature applications are also introduced and the pivotal role of calcium, strontium, silicon, and rare earths in improving high temperature properties is highlighted. Finally, a brief description of upcoming high strength magnesium-based bulk metallic glass materials is presented.

3.1. INTRODUCTION

The addition of alloying elements in pure magnesium helps to alter its properties. Magnesium is chemically active and can react with other metallic alloying elements to form intermetallic compounds. In most of the magnesium alloys, the presence of intermetallic phases can be observed. These phases aid to influence the microstructure, and hence, affect the mechanical properties of the magnesium alloy. Solid solution hardening and/or

---

Magnesium, Magnesium Alloys, & Magnesium Composites, by Manoj Gupta and Nai Mui Ling, Sharon
© 2010 John Wiley & Sons, Inc.

39

precipitation hardening are the key mechanisms to enhance the mechanical performance of the magnesium-based materials. In the following Section 3.1.1, the effects of addition of metallic elements on magnesium are discussed in alphabetical order.

3.1.1. Effects of Addition of Metallic Elements on Magnesium

3.1.1.1. Aluminum. Aluminum is one of the most commonly used alloying elements in magnesium, as it has the most favorable influence on magnesium. Addition of aluminum results in the enhancement of hardness and strength. It also improves castability. The alloy in excess of 6 wt% of aluminum can be heat treated [1].

3.1.1.2. Beryllium. Beryllium is only added to the melt in small quantities (<30 ppm). Addition of beryllium reduces the surface melt oxidation significantly during the casting, melting, and welding processes. It can also result in grain coarsening [1].

3.1.1.3. Calcium. The addition of calcium can assist in grain refinement and creep resistance. It can also enhance corrosion resistance, thermal and mechanical properties of magnesium alloys [2, 3]. Calcium when incorporated to casting alloys, reduces melt oxidation and also oxidation during heat treatment process. Its presence also results in better rollability of magnesium sheet. However, if its amount exceeds 0.3 wt%, the sheet will be prone to cracks during welding [1].

3.1.1.4. Cerium. The addition of cerium (only 0.2% Ce) results in significant increase in elongation in magnesium. Mishra et al. [4] reported that the presence of cerium in magnesium altered the texture of the extruded rods during recrystallization in a way that enhances plastic deformation capability. In the Mg–Ce material, both twinning and slip activities occur in comparison with the Mg material, whereby only twinning dominates. In addition, it was observed that addition of Ce reduces the yield strength and increases the work hardening rate that delays the onset of instability [4].

3.1.1.5. Copper. Copper has limited solid solubility in magnesium [5]. Copper reacts with magnesium to form Mg_2Cu intermetallics. Studies have indicated that the addition of copper in magnesium assists in increasing the room temperature [6–8] and high temperature strength [1]; however, the ductility is compromised [6–8]. It may also be noted that copper addition can unfavorably influence the corrosion resistance [1].

3.1.1.6. Iron. Iron is a harmful addition in magnesium alloys, and even in small quantities, its presence is detrimental to the corrosion resistance. A total of 0.005% of iron content is the upper limit allowed for best protection against corrosion [1].

3.1.1.7. Lithium. Lithium has relatively high solid solubility in magnesium (17.0 at.%, 5.5 wt%) [5]. The addition of lithium can decrease the density of alloys below that of unalloyed magnesium, owing to its low relative density of 0.54. Its addition leads to decrease in strength and increase in ductility [1].

3.1.1.8. Manganese. The addition of manganese enhances the saltwater corrosion resistance of Mg–Al and Mg–Al–Zn alloys. The low solubility of manganese in magnesium limits the amount of manganese addition in magnesium. Manganese is usually incorporated with other alloying elements like aluminum [1].

3.1.1.9. Molybdenum. Molybdenum does not alloy and interact with magnesium [5]. Wong et al. [9] reported that increasing weight percent of molybdenum (0.7, 2.0, and 3.6 wt%) in magnesium resulted in an improvement in hardness, elastic modulus and ductility, while strength was marginally reduced.

3.1.1.10. Nickel. Nickel has limited solid solubility in magnesium [5]. Hassan et al. [10] reported that the incorporation of nickel in magnesium led to the formation of Mg_2Ni intermetallics and an increase in room temperature strength (in terms of increase in 0.2% yield strength and ultimate tensile strength); however, a decrease in ductility was observed. Nickel additions in magnesium alloys even in very minute amounts have adverse effect on the corrosion resistance [1].

3.1.1.11. Rare Earth Metals (RE). Rare earths are added to increase the high temperature strength, creep resistance, and corrosion resistance. However, as rare earths are costly, they are primarily used in high-tech alloys. Their presence also assists in reducing the freezing range of the alloys, which results in less casting porosity and weld cracking [1].

3.1.1.11.1. NEODYMIUM. Addition of neodymium enhances the strength of magnesium. This is attributed to the solubility limits of neodymium and the formation of stable precipitates within the grain structure and at grain boundaries. Elektron 21, which is developed by Magnesium Elektron for aerospace and motorsport applications, contains neodymium, heavy rare earth (gadolinium), zinc, and zirconium [11].

3.1.1.12. Silicon. Silicon can increase the fluidity of molten alloys. However, if it is used together with iron, the corrosion resistance will be compromised [1].

3.1.1.13. Silver. Silver used in conjunction with rare earths improves the high temperature strength and creep resistance [1].

3.1.1.14. Strontium. Strontium is normally added to magnesium together with other major alloying elements. It has been reported that incorporation of strontium enhances the creep performance but exhibit no significant impact on yield and ultimate tensile strengths [12].

3.1.1.15. Thorium. The addition of thorium leads to an improvement in creep strength up to 370°C. For alloys containing zinc, the addition of thorium also enhances the weldability. However, due to the radioactive nature of thorium, it is replaced by other alloying elements [1].

3.1.1.16. Tin. Tin together with aluminum in magnesium improves ductility. It also assists in reducing cracking tendency during forging [1].

3.1.1.17. Titanium. There is very limited mutual solubility of titanium and magnesium in each other [5]. Hassan et al. [13] reported that respective additions of 2.2 and 4.0 vol.% of titanium in magnesium led to an increase in 0.2% yield strength and ductility. No intermetallic compound was formed.

3.1.1.18. Yttrium. Yttrium has a relatively high solid solubility in magnesium (~12.6 wt%, 3.75 at.%) [5]. It is incorporated with other rare earth metals to enhance the high temperature strength and creep performance [1].

3.1.1.19. Zinc. Zinc is one of the most commonly used and effective alloying elements in magnesium. It is usually used in conjunction with aluminum to increase the strength without reducing ductility. The increase in strength to comparable levels is not achievable if only aluminum content is increased [14]. Moreover, zinc incorporated into magnesium alloy with nickel and iron impurities can also assist to improve the corrosion resistance [1].

3.1.1.20. Zirconium. Zirconium can function as an excellent grain refiner when incorporated into alloys containing zinc, thorium, rare earths, or a combination of these elements. However, it cannot be used with aluminum or manganese because of the formation of stable compounds with these alloying elements. Furthermore, it forms stable compounds with iron, carbon, oxygen, and hydrogen that are present in the melt [1].

3.1.2. Classifications of Magnesium Alloys

3.1.2.1. Alloy Designations. Table 3.1 shows the abbreviation letters for the alloying elements that are commonly used to designate the magnesium alloys, as given by the American Society for Testing and Materials (ASTM B275) [15]. Each alloy is labeled by letters that indicate the main alloying elements, followed by figures that represent the percentages of these elements.

The designation of a typical magnesium alloy consists of three parts. For part 1, the initial alphabet of two main alloying elements makes two abbreviation letters (refer to Table 3.1), which represent the two main alloying elements arranged in order of decreasing percentage. If the percentages of the alloying elements are equal, the letters are arranged alphabetically. For part 2, the amounts (in weight percentage terms) of the two main alloying elements are stated. It consists of two whole numbers, which corresponds to the two alphabets. For part 3, it distinguishes between the

TABLE 3.1. ASTM designation system of magnesium alloys [1, 16].

Alloying element	Abbreviation letter
Aluminum	A
Bismuth	B
Copper	C
Cadmium	D
Rare earth metals	E
Iron	F
Thorium	H
Zirconium	K
Lithium	L
Manganese	M
Nickel	N
Lead	P
Silver	Q
Chromium	R
Silicon	S
Tin	T
Yttrium	W
Antimony	Y
Zinc	Z

different alloys with the same percentages of the two main alloying elements. It is made up of a letter of the alphabet assigned in order as compositions become standard, namely:

A	First compositions, registered with ASTM
B	Second compositions, registered with ASTM
C	Third compositions, registered with ASTM
D	High purity, registered with ASTM
E	High corrosion resistance, registered with ASTM
X	Experimental alloy, not registered with ASTM

For example, considering magnesium alloy AZ91C:

AZ	Indicates that aluminum and zinc are the two main alloying elements
91	Indicates the percentages of aluminum and zinc (9 and 1, respectively), which are rounded-off to whole numbers
C	Indicates the third specific composition registered having this nominal composition

3.1.2.2. Temper Designations. Other fabrication conditions can also be included as shown in Table 3.2. Temper designations are in accordance with ASTM B296-03 [17]. A dash is used to separate the alloy designation from the temper designation, such as AZ91C-T4. Table 3.3 lists some of the commonly used magnesium alloys.

TABLE 3.2. Temper designations for magnesium alloys [1].

General Divisions	
F	As fabricated
O	Annealed, recrystallized (wrought products only)
H	Strain hardened
T	Thermally treated to produce stable tempers other than F, O, or H
W	Solution heat treated (unstable temper)
Subdivisions of "H"	
H1, plus one or more digits	Strain hardened only
H2, plus one or more digits	Strain hardened and then partially annealed
H3, plus one or more digits	Strain hardened and then stabilized
Subdivisions of "T"	
T1	Cooled from an elevated temperature shaping process and naturally aged
T2	Annealed (cast products only)
T3	Solution heat treated and cold worked
T4	Solution heat treated and naturally aged to a substantially stable condition
T5	Cooled from an elevated temperature shaping process and artificially aged
T6	Solution heat treated and artificially aged
T7	Solution heat treated and stabilized
T8	Solution heat treated, cold worked, and artificially aged
T9	Solution heat treated, artificially aged, and cold worked
T10	Cooled from an elevated temperature shaping process, artificially aged, and cold worked

3.2. CASTING ALLOYS

Magnesium casting alloys are produced by a range of casting processes in foundries. The general characteristics, physical, and mechanical properties of the casting alloys are presented in the following sections.

3.2.1. Characteristics of Casting Alloys

Tables 3.4 and 3.5 list the characteristics and uses of casting alloys.

3.2.2. Physical Properties of Casting Alloys

Refer to Table 3.6 and Figures 3.1 and 3.2.

TABLE 3.3. List of commonly used magnesium alloys [18].

(a) Aluminum as Main Alloying Element

Designation	Al	Fe (max.)	Mn	Ni (max.)	RE	Si	Zn
AJ52A	5	—	0.38	—	2.0 (Strontium)	—	0.2
AJ62A	6	—	0.38	—	2.5 (Strontium)	—	0.2
AM50A	4.9	0.004	0.32	0.002	—	—	0.22
AM60B	6.0	0.005	0.42	0.002	—	—	0.22 (max.)
AS41B	4.2	0.0035	0.52	0.002	—	1.0	0.12
AZ31B	3	0.005	0.6	0.005	—	—	1.0
AZ61A	6.5	0.005	0.33	0.005	—	—	0.9
AZ80A	8.5	0.005	0.31	0.005	—	—	0.5
AZ81A	7.6	—	0.24	—	—	—	0.7
AZ91D	9	0.005	0.33	0.002	—	—	0.7
AZ91E	9	0.005	0.26	0.001	—	—	0.7

(b) Other Elements as Main Alloying Element

Designation	Ag	Fe (max.)	Mn	Ni (max.)	RE	Zn	Zr
EZ33A	—	—	—	—	3.2	2.5	0.7
K1A	—	—	—	—	—	—	0.7
M1A	—	—	1.6	—	—	—	—
QE22A	2.5	—	—	—	2.2	—	0.7
WE43A	—	0.01	0.15	0.005	4 (Yttrium)	0.2	0.7
WE54A	—	—	0.15	0.005	5.1 (Yttrium)	—	0.7
ZE41A	—	—	1.5	—	1.2	4.2	0.7
ZE63A	—	—	—	—	2.6	5.8	0.7
ZK40A	—	—	—	—	—	4.0	0.7
ZK60A	—	—	—	—	—	5.5	0.7

TABLE 3.4. Characteristics and uses of die casting alloys [1].

Alloy	Temper	Characteristics and Uses
AE42	F	Good creep resistance up to 150°C and good strength
AM20	F	High impact strength and ductility
AM50A	F	Excellent energy absorbing properties and ductility; used for production of automotive wheels
AM60A	F	Similar to AM50A-F but slightly better strength; greater ductility and toughness than the AZ91 alloys; only used where good saltwater corrosion resistance is not needed
AM60B	F	Similar to AM50A-F but slightly better strength; greater ductility and toughness than the AZ91 alloys
AS21	F	Similar to AE42-F
AS41A	F	Similar to AS21-F but lower ductility and creep resistance, better castability and strength; used in automotive structural die casting parts exposed to service temperatures up to 175°C
AZ91A, B, D	F	Good strength and excellent castability; AZ91A and AZ91B are used as secondary metal to lower the cost of the alloy and are used when maximum corrosion resistance is not necessary; AZ91D is a high purity alloy with excellent corrosion resistance and is the most commonly used magnesium die casting alloy

TABLE 3.5. Characteristics and uses of sand and permanent mold casting alloys [1].

Alloy	Temper	Characteristics and Uses
AZ63A	T6	Good room temperature strength, ductility, and toughness; used for commercial and military structural parts
AZ81A	T4	Good castability, tough, low microshrinkage tendency and pressure tight; used for commercial and military structural parts
AZ91C	T6	General-purpose alloy and moderate strength; used when maximum corrosion resistance is not needed
AZ91E	T6	General-purpose alloy, moderate strength, and excellent corrosion resistance
AZ92A	T6	High tensile strength and good yield strength; used in commercial and military pressure-tight sand and permanent-mold castings
EQ21A	T6	Excellent short-time high temperature mechanical properties, pressure tight and weldable; used in commercial and military sand and permanent-mold castings
EZ33A	T5	Good damping capacity, castability, and creep resistance up to 245°C; used in commercial and military pressure-tight sand and permanent-mold castings; also used in applications that need good strength properties up to 260°C
HK31A	T6	Good castability, creep resistance up to 350°C and pressure tight
HZ32A	T5	Good castability, pressure tight, equal or improved creep resistance than HK31A-T6 up to 260°C
K1A	F	Good damping capacity
QE22A	T6	High yield strength up to 200°C and good castability; excellent short-time high temperature mechanical properties and weldable; used in commercial and military pressure-tight sand and permanent-mold castings
QH21A	T6	Creep resistance and high yield strength up to 250°C, pressure-tight and good castability
WE43A	T6	Good corrosion resistance and high strength (at room and high temperature up to 290°C); pressure tight and weldable; used in military and aerospace sand castings
WE54A	T6	Similar to WE43A-T6, gradually loses ductility when exposed up to 150°C; pressure tight and weldable
ZC63A	T6	Stronger and more castable than AZ91C; used in pressure-tight applications and weldable
ZE41A	T5	Enhanced castability over ZK51A and medium strength high temperature alloy; used in commercial and aerospace sand castings, and pressure-tight applications
ZE63A	T6	Strong, thin-walled, and pore-free casting; Excellent castability and pressure tight; used in aerospace and military sand and investment castings
ZH62A	T5	High room temperature yield strength
ZK51A	T5	Good ductility and room temperature strength; used for highly stressed parts that are simple in design and small
ZK61A	T5	Similar to ZK51A-T5 but with better yield strength; used for highly stressed, simple-designed aerospace, and military castings of uniform cross –section; high in cost; complicated castings are prone to microporosity and cracking due to shrinkage
ZK61A	T6	Similar to ZK61A-T5 but with better yield strength

TABLE 3.6. Physical properties of magnesium casting alloys [1].

Alloy	Density, at 20°C (g/cm^3)	Linear CTE, at 20–100 °C (μm/m K)	Specific Heat, at 20°C (kJ/kg K)	Latent Heat of Fusion (kJ/kg)	Thermal Conductivity, at 20°C (W/m K)	Electrical Resistivity, at 20°C (nΩ m)
AM50A	1.77	26.0	1.02	370	65 [F]	130
AM60A,B	1.80	26.0	1.00	370	61 [F]	—
AM100A	1.81	25.0	—	—	58.3 [T6]	124 [T6]
AS41A,B	1.776	26.1	1.01	413	68 [F]	—
AZ63A	1.82	27.2	—	373	59.2 [F]	122 [F]
AZ81A	1.80	27.2	—	—	51.1	128 [F], 150 [T4]
AZ91A-E	1.81	27.2	0.8	373	See Figure 3.1	See Figure 3.2
AZ92A	1.83	27.2	—	373	See Figure 3.1	See Figure 3.2
EQ21A	1.81	26.7	—	374	113	68.5
EZ33A	1.80	26.8	1.04	373	See Figure 3.1	See Figure 3.2
K1A	1.74	27.0		343–360	122	See Figure 3.2
QE22A	1.82	26.7	1.00	373	113	See Figure 3.2
WE43A	1.84	—	0.966		51.3	—
WE54A	1.85	—	—	—	52	173
ZC63A	1.87	27.0	0.960	—	122	54
ZE41A	1.84	—	—	—	See Fig. 3.1	See Fig. 3.2
ZE63A	1.87	27.1	0.960	—	109	56
ZK51A	1.83	27.1	1.02	318	110	62
ZK61A	1.83	27.0	—	—	—	—

3.2.3. Mechanical Properties of Casting Alloys

Refer to Tables 3.7–3.11.

3.3. WROUGHT ALLOYS

Wrought products are fabricated as extrusions, forgings, plate, and sheet. Wrought magnesium alloys as compared with their casting counterparts have better microstructural homogeneity and generally enhanced mechanical properties. This allows them to be used in a wide spectrum of applications [20, 21]. Their improved mechanical properties are attributed to their thermomechanical method of production.

3.3.1. Characteristics of Wrought Alloys

Refer to Tables 3.12–3.14.

3.3.2. Mechanical Properties of Wrought Alloys

Refer to Tables 3.15–3.17.

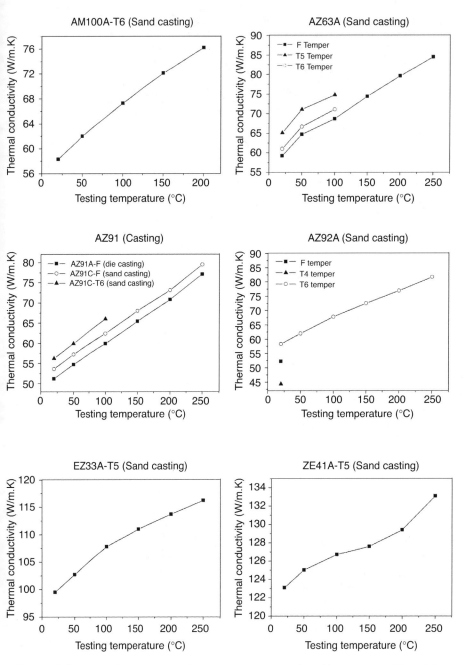

Figure 3.1. Thermal conductivity of magnesium casting alloys (data extracted from [1]).

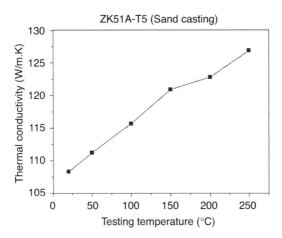

Figure 3.1. (*Continued*).

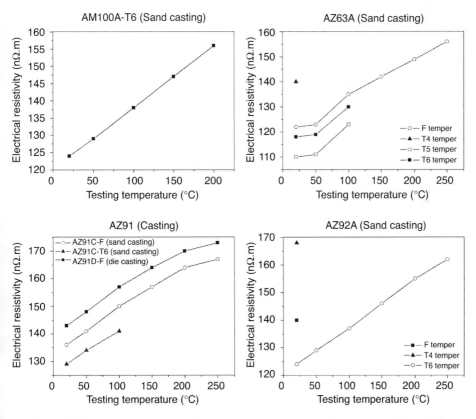

Figure 3.2. Electrical resistivity of magnesium casting alloys (data extracted from [1]).

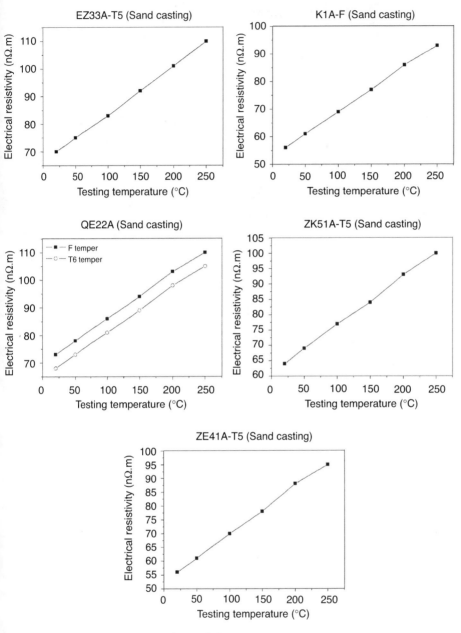

Figure 3.2. (*Continued*).

TABLE 3.7. Room temperature mechanical properties of die casting alloys [19].

Alloy	AE42	AM20	AM50	AM60	AS21	AS41	AZ91
Yield strength (MPa)	145	90	125	130	120	140	160
Tensile strength (MPa)	230	210	230	240	220	240	250
Elongation (%)	11	20	15	13	13	15	7
Hardness (HB)	60	45	60	65	55	60	70

TABLE 3.8. Elevated temperatures tensile properties of die casting alloys [18].

Alloy	ACM522		AJ52X		AJ62X	
Testing Temperature (°C)	20	175	20	175	20	175
Yield strength (MPa)	158	132	161	100	143	103
Tensile strength (MPa)	200	152	228	141	240	143
Elongation (%)	4	9	13	18	7	19

TABLE 3.9. Room temperature tensile properties of investment casting alloys [18].

Alloy	Temper	Yield Strength (MPa)	Tensile Strength (MPa)	Elongation (%)
AZ91E	F	100	165	2
	T4	100	275	12
	T5	100	180	3
	T7	140	275	5
EZ33A	T5	110	255	4
K1A	F	60	175	20
QE22A	T6	185	260	4

TABLE 3.10. Elevated temperature tensile properties of sand casting alloys [1].

Alloy	Temper	Testing Temperature (°C)	Yield Strength (MPa)	Tensile Strength (MPa)	Elongation (%)
AM100A	T4	93	—	235	1.5
		150	—	160	9.0
		260	—	83	22.0
AM100A	T6	150	62	165	4.0
		200	45	115	25.0
		260	28	83	45.0
AZ63A	F	65	—	210	3.0
		93	—	208	4.5
		120	—	191	7.5

(Continued)

TABLE 3.10. (*Continued*).

Alloy	Temper	Testing Temperature (°C)	Yield Strength (MPa)	Tensile Strength (MPa)	Elongation (%)
		150	—	166	20.5
		200	—	105	50.5
		260	—	71	38.0
AZ63A	T4	65	—	253	9.0
		93	—	236	7.0
		120	—	207	9.0
		150	—	154	33.2
		200	—	101	38.0
		260	—	75	26.0
AZ63A	T6	93	119	248	11.0
		120	114	223	11.0
		150	103	169	15.0
		200	83	121	17.0
		260	61	83	15.0
AZ81A	T4	93	83	260	20.0
		150	80	190	24.5
		200	76	140	29.0
		260	72	97	35.0
AZ91C	T6	150	97	185	40
		200	83	115	40
EZ33A	T5	150	97	150	10
		200	76	145	20
		260	69	125	31
		315	55	83	50
K1A	F	93	48	115	30
		200	34	55	71
		315	14	28	78
QE22A	T6	100	193	235	—
		200	166	193	—
		300	69	83	—
ZE41A	T5	93	138	193	8
		150	130	172	12
		200	114	141	31
		260	88	106	40
		315	69	82	45
ZE63A	T6	100	131	235	—
		150	111	187	—
		200	97	131	—
ZK51A	T5	95	145	205	12
		150	115	160	14
		205	90	115	17
		260	62	83	16
		315	41	55	16

TABLE 3.11. Room temperature tensile properties of sand and permanent mould casting alloys [18].

Alloy	Temper	Yield Strength (MPa)	Tensile Strength (MPa)	Elongation (%)
AZ81A	T4	85	276	15
AZ91E	F	95	165	3
	T4	85	275	14
	T6	195	275	6
EZ33A	T5	105	160	3
K1A	F	51	185	20
QE22A	T6	205	275	4
WE43A	T6	190	235	4
WE54A	T6	195	270	4
ZE63A	T6	190	295	7

TABLE 3.12. Characteristics of magnesium alloys produced as extruded shapes and bars [1].

Alloy	Temper	Characteristics
AZ10A	F	Moderate strength and low cost
AZ31B, C	F	Medium strength
AZ61A	F	Good strength and reasonable cost
AZ80A	T5	Higher strength than AZ61A-F
M1A	F	Good damping capacity and corrosion resistance, low to medium strength
ZC71	T6	Good ductility and strength; medium cost
ZK21A	F	Reasonable strength and good weldability
ZK31	T5	High strength and good weldability
ZK40A	T5	High strength, not weldable, and better extrudability than ZK60A
ZK60A	T5	Not weldable and high strength
ZM21	F	Medium strength, good damping capacity and formability

TABLE 3.13. Characteristics of magnesium alloys produced as plate and sheet [1].

Alloy	Temper	Characteristics
AZ31B	H24	Medium strength
ZM21	O	Good damping capacity and formability
ZM21	H24	Medium strength

TABLE 3.14. Characteristics and uses of magnesium alloys produced as forgings [1].

Alloy	Temper	Characteristics
AZ31B	F	Medium strength, good forgeability but not often used
AZ61A	F	Better strength than AZ31B-F
AZ80A	T5	Better strength than AZ31B-F and AZ61A-F
AZ80A	T6	Better creep resistance than AZ80A-T5
M1A	F	Low to medium strength, good corrosion resistance but not often used
ZK31	T5	Medium weldability and high strength
ZK60A	T5	Comparable strength with AZ80A-T5 but better ductility
ZK61	T5	Similar to AZ60A-T5
ZM21	F	Medium strength, good forgeability and damping capacity

TABLE 3.15. Room temperature mechanical properties of extruded products [1, 18].

Alloy	Temper	Yield Strength (MPa)		Tensile Strength (MPa)	Elongation (%)	Shear Strength (MPa)	Hardness (HRB)
		Compressive	Tensile				
AZ31B	F	95	200	260	15	130	49
AZ61A	F	130	230	310	16	140	60
AZ80A	F	140	250	340	11	—	—
	T5	240	275	380	7	165	82
HM31A	F	185	230	290	10	150	—
M1A	F	83	180	255	12	125	44
ZC71	F	—	340	360	5	—	70–80
ZK21A	F	135	195	260	4	—	—
ZK40A	T5	140	255	275	4	—	—
ZK60A	F	185	250	340	14	—	—
	T5	250	305	365	11	180	88

TABLE 3.16. Room temperature mechanical properties of plates and sheets [1, 18].

Alloy	Temper	Yield Strength (MPa)		Tensile Strength (MPa)	Elongation (%)	Shear Strength (MPa)	Hardness (HRB)
		Compressive	Tensile				
AZ31B	H24	180	220	290	15	160	73
	O	110	150	255	21	—	—
HK31A	H24	160	200	255	9	140	68
HM21A	T8	130	170	235	11	128	—

TABLE 3.17. Room temperature mechanical properties of forged products [1, 18].

Alloy	Temper	Yield Strength (MPa)		Tensile Strength (MPa)	Elongation (%)
		Compressive	Tensile		
AZ31B	F	85	195	260	9
AZ61A	F	115	180	195	12
AZ80A	F	170	215	315	8
	T5	195	235	345	6
	T6	185	250	345	5
ZK60A	T5	195	205	305	16
	T6	170	270	325	11

3.4. MAGNESIUM ELEKTRON SERIES ALLOYS

3.4.1. Magnesium Elektron Casting Alloys [22]

Since the 1930s, Magnesium Elektron has been supplying magnesium casting alloys. A wide range of magnesium alloy systems with enhanced mechanical properties at room and elevated temperatures up to 300°C have been developed. The following section lists the characteristics, uses, and properties of the casting alloys from Magnesium Elektron [22] Tables 3.18–3.20.

TABLE 3.18. Characteristics and uses of Magnesium Elektron casting alloys [22].

Alloy	Compositions	Characteristics and Uses
Elektron 21	Zn (0.2–0.5%) Nd (2.6–3.1%) Gd (1.0–1.7%) Zr (saturated) Mg (balance)	High strength, excellent corrosion resistance and castability; used in aerospace and motor sport applications and used at temperatures up to 200°C
Elektron AZ91E	Al (8.1–9.3%) Zn (0.4–1.0%) Mn (0.17–0.35%) Mg (balance)	Good castability and excellent corrosion resistance; gravity sand casting alloy; used in aerospace casting applications where pressure tightness and high temperature requirements are not needed
Elektron EQ21	Ag (1.3–1.7%) Rare earths (1.5–3.0%) Cu (0.05–0.10%) Zr (0.4–1.0%) Mg (balance)	A cheaper alternative to Elektron MSR-B due to the lower Ag content; high strength; good room and high temperature properties; weldable and pressure tight; used at temperatures up to 200°C; used in military, aerospace, and automotive applications where good retention of properties at high temperatures is desired

(Continued)

TABLE 3.18. (*Continued*).

Alloy	Compositions	Characteristics and Uses
Elektron MSR-B	Ag (2.0–3.0%) Rare earths (2.0–3.0%) Zr (0.4–1.0%) Mg (balance)	High strength; good room and high temperature properties; weldable and pressure tight; used at temperatures up to 200°C; used in military, aerospace, and automotive applications where good retention of properties at high temperatures is desired
Elektron RZ5 (ZE41)	Zn (3.5–5.0%) Rare earths (0.8–1.7%) Zr (0.4–1.0%) Mg (balance)	High strength, excellent castability, pressure tight, and weldable; best for high-integrity castings used in room and up to 150°C; used in T5 condition and in automotive, aerospace, military, and electronic applications; examples of specific applications include motorcycle wheels, aircraft parts and engines, power tools, computer parts, video cameras, helicopter gearboxes, and performance car parts
Elektron WE43	Y (3.7–4.3%) Rare earths (2.4–4.4%) Zr (min 0.4%) Mg (balance)	Excellent corrosion resistance and good mechanical properties at high temperatures; excellent castability and pressure tight; for use at temperatures up to 300°C and in applications that require stability for long-term exposure up to 250°C; examples of applications include helicopter transmissions, missiles, racing cars, performance cars, and aero engines
Elektron WE54	Y (4.75–5.5%) *Heavy rare earths (1.0–2.0%) Nd (1.5–2.0%) Zr (min 0.4%) Mg (balance) *Contains: Yb, Er, Dy, and Gd	High strength at high temperatures; excellent corrosion resistance, pressure tight, and good castability; for use at temperatures up to 300°C and in applications such as transmissions, missiles, performance cars, and power systems
Elektron ZRE1 (EZ33)	Zn (2.0–3.0%) Rare earths (2.5–4.0%) Zr (0.4–1.0%) Mg (balance)	Creep resistant up to 250°C; excellent castability, pressure tight, and weldable; used in T5 condition and for low-stressed intricate casting parts that are operating at temperatures where creep resistance is essential

3.4.2. Wrought Magnesium Elektron Alloys [22]

Wrought magnesium alloys are produced in a wide range of products such as magnesium bars, billets, sheets, and sections. The following section lists the characteristics, uses, and properties of the wrought magnesium alloys from Magnesium Elektron [22] (Figures 3.3–3.6, Tables 3.21 and 3.22).

TABLE 3.19. Properties of Magnesium Elektron casting alloys [22].

Alloy	Elektron 21	AZ91E	EQ21	MSR-B	RZ5 (ZE41)	WE43	WE54	ZRE1 (EZ33)
Specific gravity	1.82	1.81	1.81	1.82	1.84	1.84	1.85	1.80
CTE ($\times 10^{-6}$ K^{-1})	25.3	27	26.7	26.7	27.1	26.7	24.6	26.8
Thermal conductivity (W m^{-1} K^{-1})	116	84	113	113	109	51.3	52	100
Specific heat (J kg^{-1} K^{-1})	1086	1000	1000	1000	960	966	960	1040
Electrical resistivity (nΩ m)	94.6	141	68.5	68	68	148	173	73
Melting range (°C)	545–640	470–595	540–640	550–640	510–640	540–640	545–640	545–640
Poisson's ratio	0.27	0.20	0.30	0.40	0.35	0.27	0.30	0.33
Damping index	—	0.20	0.22	0.40	1.00	0.09	0.17	1.89

TABLE 3.20. Ambient temperature mechanical properties of Magnesium Elektron casting alloys [22].

Alloy	Elektron 21	AZ91E	EQ21	MSR-B	RZ5 (ZE41)	WE43	WE54	ZRE1 (EZ33)
Elastic modulus (GPa)	44	44	44	44	44	44	44	44
Hardness	65–75 (Brinell)	75 (Brinell)	80–105 (Vickers)	80–105 (Vickers)	55–70 (Brinell)	85–105 (Vickers)	80–90 (Brinell)	50–60 (Brinell)
Tensile 0.2% Proof stress (MPa)	170	125 (T4) 170 (T6)	195	205	148	180	205	110
Tensile strength (MPa)	280	260 (T4) 270 (T6)	261	266	218	250	280	160
Elongation (%)	5.0	9.0 (T4) 4.5 (T6)	4.0	4.0	4.5	7.0	4.0	3.0
Compressive 0.2% Proof stress (MPa)	168	130	165–200	165–200	130–150	187	165–175	85–120
Ultimate compressive strength (MPa)	367	400	310–385	310–385	330–365	323	410	275–340
Ultimate shear strength (MPa)	172	140	152	152	138	162	150	138
Fracture toughness, K_{IC} (MPa m$^{1/2}$)	—	13.2	16.4	14.9	15.1–16.3	15.9	14.3	—

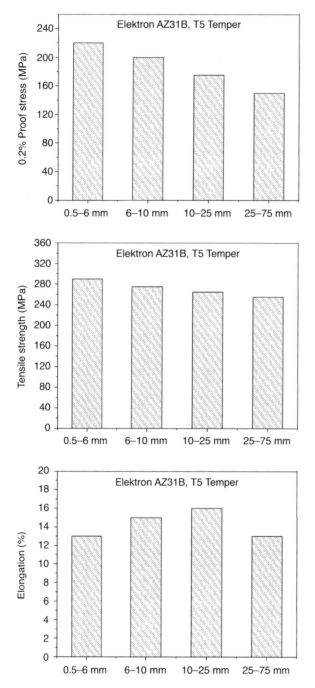

Figure 3.3. Effect of dimensional variation on tensile properties of wrought Elektron AZ31B alloy (data extracted from [22]).

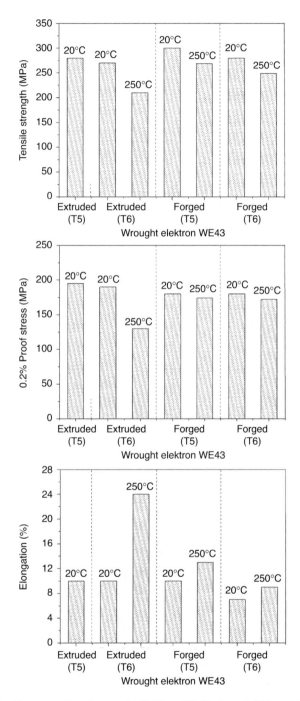

Figure 3.4. Tensile properties of wrought Elektron WE43 alloy at different testing temperatures (data extracted from [22]).

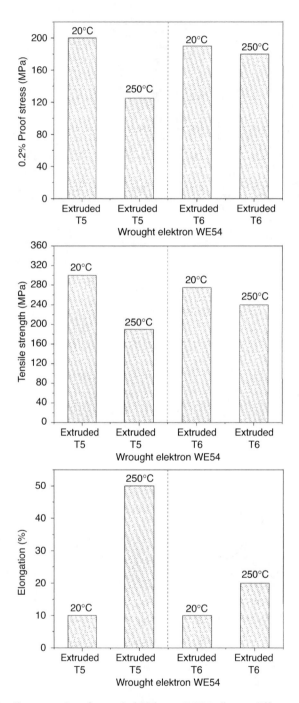

Figure 3.5. Tensile properties of extruded Elektron WE54 alloy at different testing temperatures (data extracted from [22]).

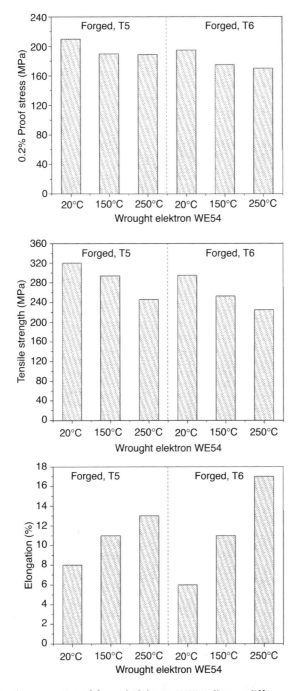

Figure 3.6. Tensile properties of forged Elektron WE54 alloy at different testing temperatures (data extracted from [22]).

TABLE 3.21. Characteristics and uses of wrought Magnesium Elektron alloys [22].

Alloy	Compositions	Characteristics and Uses
Elektron AZ31B Coil, plate, and sheet	Al (2.5–3.5%) Zn (0.7–1.3%) Mn (0.2–1.0%) Mg (balance)	Good room temperature strength, ductility, corrosion resistance, and excellent weldability; Nonmagnetic with high electrical and thermal conductivity; used for RFI and EMI shielding in computer and electronics industries AZ31B plate and sheet: used in applications requiring medium strength at temperatures below 150°C; some applications include cell phone, camera, casing of notebook and computer, and aircraft fuselage; superplastic forming of AZ31B sheet at high temperatures can be used for complicated automotive parts
Elektron wrought WE43	Y (3.7–4.3%) *Rare earths (2.4–4.4%) Zr (min 0.4%) Mg (balance) *Contains 2.0–2.5% Nd and remainder are Yb, Er, Dy, and Gd	High strength and stable under long-term exposure at temperatures up to 250°C; more isotropic mechanical properties when compared with conventional wrought magnesium alloys; used at temperatures up to 300°C; excellent corrosion resistance and excellent retention of properties at high temperature; used in aircraft, performance cars, and aero engines
Elektron wrought WE54	Y (4.75–5.5%) *Rare earths (3.0–4.0%) Zr (min 0.4%) Mg (balance) *Contains: 1.5–2.0% Nd, remainder are Yb, Er, Dy, and Gd	High strength and more isotropic mechanical properties when compared with conventional magnesium alloys; used at temperatures up to 300°C; excellent corrosion resistance and excellent retention of properties at high temperature; used in aircraft, performance cars, and aero engines
Elektron wrought ZK60A	Zn (4.8–6.2%) Zr (min 0.45%) Mg (balance)	Not weldable by conventional methods and friction stir welding can be used; excellent resistance welding response; ZK60A forgings used in applications that require high strength and operating temperature below 150°C ZK60A-T5 has the best combination of ductility and strength at room temperature

3.5. MAGNESIUM ALLOYS FOR ELEVATED TEMPERATURE APPLICATIONS

Magnesium alloys due to their lightweight properties have found applications in the automotive industry, such as instrument panel beams, cam covers, steering wheel armatures, steering column supports, seats, transfer cases, and housings and brackets [23].

TABLE 3.22. Properties of wrought magnesium Elektron alloys [22].

Alloy	Elektron AZ31B Coil, Plate and Sheet	Elektron Wrought WE43	Elektron Wrought WE54	Elektron Wrought ZK60A
Specific gravity	1.78	1.84	1.85	1.83
CTE ($\times 10^{-6}$ K^{-1})	26.8	26.7	24.6	27.1
Thermal conductivity (W m^{-1} K^{-1})	76.9	51.3	52.0	121.0
Specific heat (J kg^{-1} K^{-1})	1040	966	960	1100
Electrical resistivity (nΩ m)	92	148	173	57
Melting range (°C)	566–632	540–640	545–640	520–635
Poisson's ratio	0.35	0.27	0.17	0.35

These components are mainly fabricated using the conventional magnesium alloys such as Mg–Al–Mn (AM alloys) and Mg–Al–Zn systems (AZ alloys) [24–26]. Although these alloys have an excellent combination of die castability, corrosion resistance and mechanical properties, they exhibit poor creep resistance at temperatures exceeding 125°C [27]. In view of the increasing need to further reduce the weight of automobiles, another growing application area of magnesium alloys is in the powertrain system (where engine pistons can operate up to 300°C, engine blocks up to 200°C and automatic transmission cases up to 175°C). These applications are also subjected to cyclic thermal and mechanical loadings and require materials with the ability to resist permanent plastic deformation due to creep. Presently, the automotive powertrain components are made of cast iron or aluminum alloy (A380).

In the past decade, much efforts have been put into the development of creep-resistant magnesium alloys for die casting applications [12,28–40]. The newly developed magnesium alloys can be categorized as follows:

   (i) Mg–Al–RE alloys
  (ii) Mg–Al–Ca alloys
 (iii) Mg–Al–Ca–RE alloys
  (iv) Mg–Zn–Al–Ca alloys
   (v) Mg–Al–Sr alloys
  (vi) Mg–Al–Si alloys
 (vii) Mg–RE–Zn alloys

3.5.1. Mg–Al–RE Alloys

Mg–Al–RE alloys are made up of at least one, and in general a mixture of, REs as the precipitate-forming alloying addition in their compositions. As REs are costly, a cheaper alternative known as misch metal is commonly used. The AE alloy series (AE41, AE42, and AE21), which contain 2–4 wt% Al, are a group of creep resistant magnesium alloys

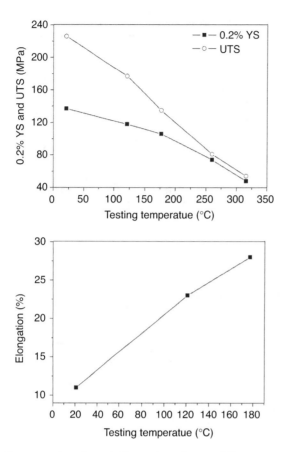

Figure 3.7. Tensile properties of AE42 die casting alloy at different testing temperatures (data extracted from [25]).

[25, 26]. Figure 3.7 shows the tensile properties of AE42 die casting alloy tested at different temperatures.

The good creep resistance property exhibited by the Mg–Al–RE alloy system can be attributed to the suppression of the formation of the β-(Mg$_{17}$Al$_{12}$) phase and the presence of Al–RE containing intermetallics (Al$_{11}$RE$_3$ and Al$_2$RE) [14, 25, 28]. Powell et al. [41] reported the findings of microstructural change in the AE42 alloy, which was subjected to creep tests at 150°C and 175°C. As shown in Table 3.23, no Mg$_{17}$Al$_{12}$ phase was observed in the as-cast alloy and at 150°C. However, at 175°C, due to thermal instability, Al$_{11}$RE$_3$ decomposed to Al$_2$RE and the excess Al reacted with Mg to form the Mg$_{17}$Al$_{12}$ intermetallic compound. Mg$_{17}$Al$_{12}$ is known to be the contributing phase for the poor creep resistance in magnesium [25, 42].

Figure 3.8 shows that AE42 alloy exhibits superior creep performance over AZ91 at high temperature. However, its creep strength is still inferior to that of the aluminum

TABLE 3.23. Amount of intermetallic compounds (wt%) present in the AE42 alloy in the as-cast and under different creep test conditions [41].

Phases	Testing Conditions		
	As-Cast, Room Temperature	150°C	175°C
Mg	97.5	97.7	97.0
$Mg_{17}Al_{12}$	0.0	0.0	0.6
$Al_{11}RE_3$	1.8	1.5	1.2
Al_2RE	0.8	0.8	1.3

alloy (A380). Figure 3.9 shows the 100 h total creep extension for AE42 alloy at 177°C increasing with increasing stress levels.

3.5.2. Mg–Al–Ca Alloys

Calcium, Ca, is a lighter and cheaper substitute to RE elements. The incorporation of Ca in Mg–Al-based alloys is also known to improve the material's creep resistance [25,26]. When the mass ratio of Ca/Al is more than ~0.8, Al_2Ca and Mg_2Ca are observed and their presence aid to increase the material's hardness. However, when this ratio is below ~0.8, only the Al_2Ca compound is formed. Ninomiya et al. [43] reported the formation of two intermetallic compounds (Al_2Ca and Mg_2Ca) along the grain boundaries and with increasing amount of Al and Ca, these precipitates increased in volume fraction.

In a study conducted by Pekguleryuz and his coworkers [44, 45], it was revealed that Mg–Al–Ca (AX) alloys with 2–6 wt% Al and 0.6–1.0 wt% Ca had creep resistance

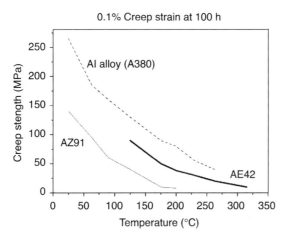

Figure 3.8. Effect of temperature on the creep strength of AE42 alloy, AZ91 alloy, and Al alloy (A380) (data extracted from [25]).

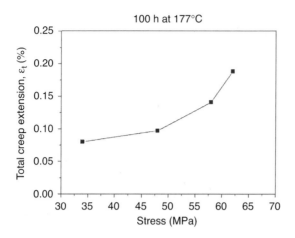

Figure 3.9. 100 h total creep extension for AE42 alloy (Mg–Al–RE) at 177°C (data extracted from [25]).

at 150°C similar to that of AE42 alloy, in addition to the good resistance to corrosion. At high temperature, $Mg_{17}Al_{12}$ intermetallic compound is suppressed while the Al_2Ca compound is thermally stable. In a 2001 patent by General Motors (GM) [46], it was revealed that when the Ca level was about 1%, the AX alloys were prone to casting defects such as hot cracking, die sticking, and cold shuts. When the Ca levels were more than 2%, these issues were reduced. Moreover, with the incorporation of ~0.1% Strontium (Sr) in Mg–Al–Ca alloys, their creep performance was enhanced.

Figures 3.10–3.12 show the tensile properties of various AX (Mg–Al–Ca), AXJ (Mg–Al–Ca–Sr), and AM50 alloys at room temperature and 175°C. The figures evidently revealed that all AX and AXJ alloys exhibited better yield strength and ultimate tensile strength when compared with that of AM50 alloy. This is attributed to the addition of Ca. It was noted that the addition of Sr did not have significant effect on the yield strength and ultimate tensile strength.

The 100 h total creep extension and secondary creep rate of AE42 (Mg–Al–RE), AX (Mg–Al–Ca), and AXJ (Mg–Al–Ca–Sr) alloys tested under different conditions are illustrated in Figures 3.13 and 3.14. Under 150°C and 175°C test conditions, it was observed that the AX and AXJ alloys yielded superior creep properties over the AE42 alloy.

3.5.3. Mg–Al–Ca–RE Alloys

MRI 153 alloy, which is a Mg–Al–Ca—RE-based alloy, is patented by Dead Sea Magnesium (Israel) and Volkswagen AG (Germany) [47, 48]. Table 3.24 summarizes the mechanical properties of MRI 153 (Mg–Al–Ca–RE) in comparison with the other magnesium alloys. The results revealed that MRI 153 alloy had superior mechanical properties and corrosion resistance over that of AE42 alloy, except for its tensile elongation

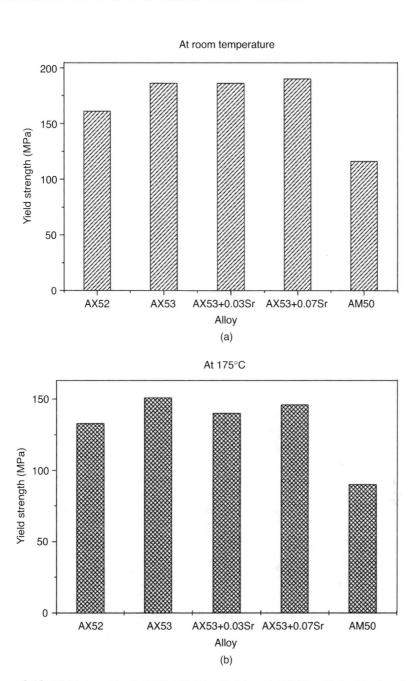

Figure 3.10. Yield strength of AM50, AX (Mg–Al–Ca), and AXJ (Mg–Al–Ca–Sr) alloys tested at (a) room temperature and (b) 175°C (data extracted from [12]).

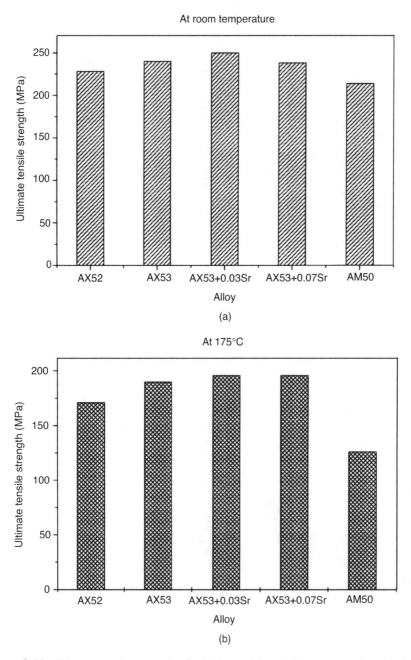

Figure 3.11. Ultimate tensile strength of AM50, AX (Mg–Al–Ca), and AXJ (Mg–Al–Ca–Sr) alloys tested at (a) room temperature and (b) 175°C (data extracted from [12]).

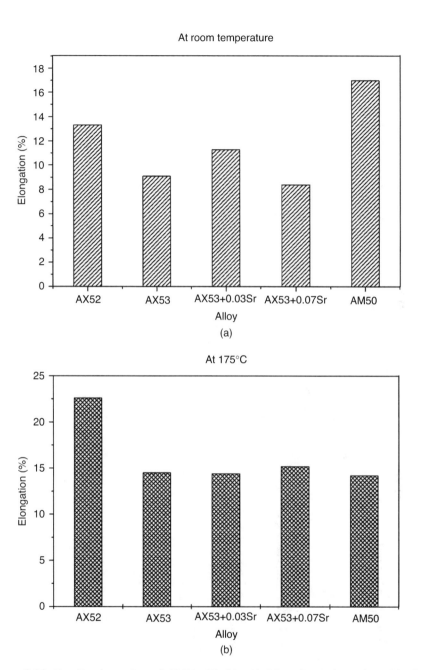

Figure 3.12. Tensile elongation of AM50, AX (Mg–Al–Ca) and AXJ (Mg–Al–Ca–Sr) alloys tested at (a) room temperature and (b) 175°C (data extracted from [12]).

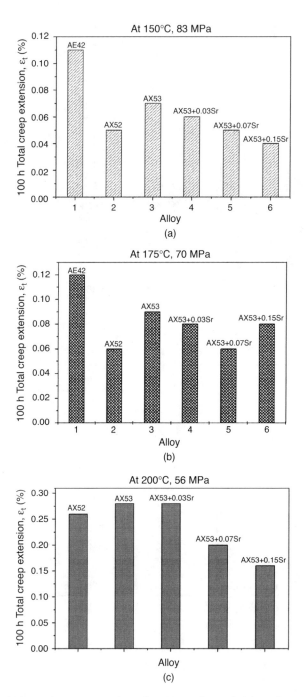

Figure 3.13. 100 h total creep extension of AE42, AX (Mg–Al–Ca), and AXJ (Mg–Al–Ca–Sr) alloys tested at (a) 150°C, (b) 175°C, and (c) 200°C (data extracted from [32]).

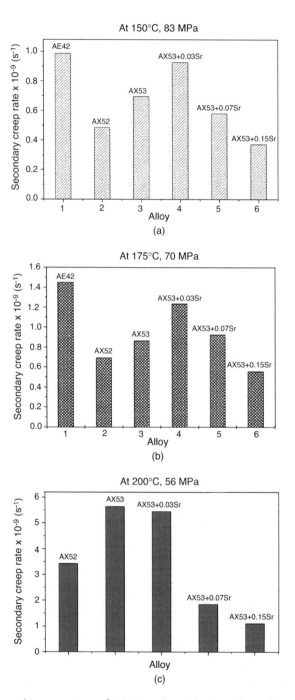

Figure 3.14. Secondary creep rate of AE42, AX (Mg–Al–Ca), and AXJ (Mg–Al–Ca–Sr) alloys tested at (a) 150°C, (b) 175°C, and (c) 200°C (data extracted from [32]).

TABLE 3.24. Mechanical properties of MRI 153 (Mg–Al–Ca–RE), AE42, and AZ91D alloys [48].

Alloy	Compressive YS (MPa)		Tensile YS (MPa)		UTS (MPa)	Elongation (%)	Fatigue Strength (MPa)	Impact Strength (J)	Corrosion Rate (mg cm^{-2} per day)
	20°C	150°C	20°C	150°C					
MRI 153	165	125	165	118	250	5	120	8	0.09
AE42	135	100	135	100	240	12	90	12	0.12
AZ91D	160	105	160	105	260	6	100	9	0.11

and impact strength. Furthermore, the mechanical properties and corrosion resistance of MRI 153 alloy are similar to that of AZ91D alloy.

3.5.4. Mg–Zn–Al–Ca Alloys

In an attempt to overcome the issues of casting defects in AX alloys, a fourth alloying element (zinc) was introduced. It was reported that the addition of 8% of Zn helped to improve the die castability of AX alloys. As shown in Figures 3.15 and 3.16, the ZAX (Mg–Zn–Al–Ca) alloys exhibit comparable, if not better, yield strength, ultimate tensile strength, and tensile elongation with that of AZ91D, at both room temperature and 150°C.

Figure 3.17 illustrates that all three ZAX alloys exhibit improved creep performance (in terms of lower total creep extension) over that of AZ91D. In comparison with AE42 alloy, ZAX8506 (Mg–8Zn–5Al–0.6Ca) and ZAX8512 (Mg–8Zn–5Al–1.2Ca) exhibited lower total creep extension values.

3.5.5. Mg–Al–Sr Alloys

The Mg–Al–Sr alloys were developed with the objective of replacing the RE alloying element in magnesium. The tensile properties of the AJ51x (Mg–5Al–1.2Sr) and AJ52x (Mg–5Al–1.8Sr) alloys are presented in Table 3.25. The results revealed that at elevated testing temperatures of 150°C and 175°C, both AJ52x and AJ51x alloys yielded improved tensile properties (in terms of yield strength (YS) and ultimate tensile strength (UTS)) over that of AE42 alloy (refer to Table 3.25). Their corrosion rate and creep property (in terms of total creep extension data) are also listed in Table 3.26. From the salt spray corrosion resistance results, it was noted that both AJ52x and AJ51x alloys outperformed the AE42 alloy. Moreover, AJ52x and AJ51x alloys also had comparable if not higher creep resistance than the AE42 alloy.

Microstructural characterization analysis conducted by Pekguleryuz et al. [49] revealed that the Sr/Al ratio contributes to the different microstructural features observed. When Sr/Al is lower than 0.3, only the Al$_4$Sr intermetallic phase was present. However, when the ratio increases, a ternary Mg–Al–Sr phase is also observed in addition of Al$_4$Sr phase. In the cases of AJ51x and AJ52x alloys, no Mg$_{17}$Al$_{12}$ intermetallic compounds

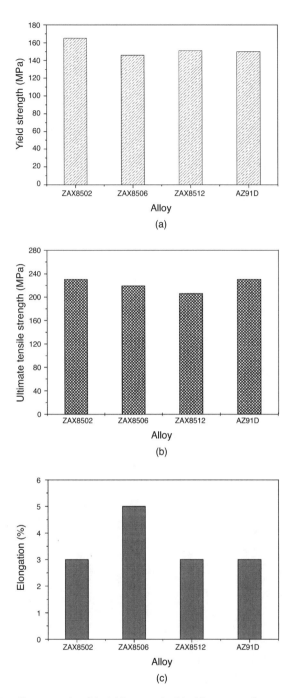

Figure 3.15. Tensile properties: (a) yield strength, (b) ultimate tensile strength, and (c) elongation of ZAX (Mg–Zn–Al–Ca) alloys at room temperature (data extracted from [30, 31]).

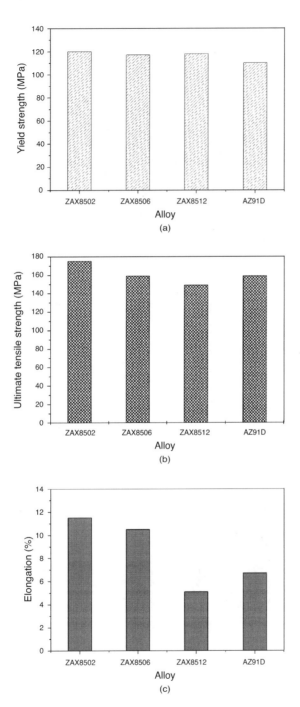

Figure 3.16. Tensile properties: (a) yield strength, (b) ultimate tensile strength, and (c) elongation of ZAX (Mg–Zn–Al–Ca) alloys at 150°C (data extracted from [30, 31]).

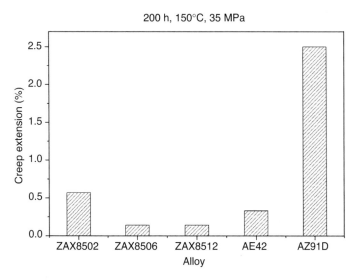

Figure 3.17. 200 h total creep extension values of various magnesium alloys subjected to 35 MPa stress and 150°C (data extracted from [30, 31]).

TABLE 3.25. Tensile properties of AJ (Mg–Al–Sr) alloys at room temperature, 150°C, and 175°C [33].

	Room Temperature			150°C			175°C		
Alloy	Tensile YS (MPa)	Tensile UTS (MPa)	Tensile Elongation (%)	Tensile YS (MPa)	Tensile UTS (MPa)	Tensile Elongation (%)	Tensile YS (MPa)	Tensile UTS (MPa)	Tensile Elongation (%)
AJ51x	138	233	8.8	102	149	16.4	97	133	21.4
AJ52x	145	202	4.0	108	164	13.6	103	148	14.8
AE42	135	226	9.2	87	142	22.5	81	121	23.1

TABLE 3.26. Corrosion rate and total creep extension (ε_t) of AJ (Mg–Al–Sr) alloys [33].

	Corrosion Rate	ε_t, 35 MPa/200 h		ε_t, 50 MPa/150°C	
Alloy	(mg cm^{-2} per day)	150°C	175°C	200 h	500 h
AJ51x	0.14	0.07	0.05	0.07	0.09
AJ52x	0.09	0.03	0.09	0.03	0.03
AE42	0.21	0.07	0.14	0.06	0.08

were found and this coupled with the addition of Al_4Sr, attribute to the alloys' superior creep resistance.

3.5.6. Mg–Al–Si Alloys

The addition of silicon results in the formation of Mg_2Si intermetallic compound. The Mg_2Si phase in Mg–Al–Si (AS) alloys has several beneficial characteristics such as high melting point (1085°C), high hardness, low density (1.9 g cm^{-3}), and low coefficient of thermal expansion (7.5×10^{-6} K^{-1}) [50]. Figure 3.18 shows the tensile properties of the AS41 (Mg–4Al–1Si) alloy under different test conditions. While Table 3.27 summarizes the mechanical properties of the AS21 (Mg–2Al–1Si) alloy.

Table 3.28 presents the corrosion rate and total creep extension data of AS41, AS21 (Mg–Al–Si), AZ91D magnesium alloys, and Al (A380) alloy. As observed, the AS21 alloy has higher corrosion rates than the AS41 and AZ91D alloys. This is due to the lower aluminum content in the AS21 alloy. In the case of AS41 alloy, it exhibits enhanced creep resistance (in terms of lower total creep extension result) over that of AZ91D under all test conditions. However, the Al alloy (A380) still exhibits superior creep resistance than the AS41 alloy.

3.5.7. Mg–RE–Zn Alloys

Magnesium Elektron Ltd (MEL) based in the United Kingdom developed a nonaluminum containing magnesium alloy, the MEZ (Mg–2.5RE–0.35Zn–0.3Mn) alloy [51]. In a study by Moreno et al. [52], it was reported that the MEZ alloy exhibit lower yield strength and tensile elongation when compared to that of the AE42 alloy. Figure 3.19 shows the effect of different testing temperatures on the tensile properties of the MEZ (Mg–2.5RE–0.35Zn–0.3Mn) and AE42 alloys. However, the MEZ alloy exhibits much better creep resistance than the AE42 alloy [29] and this is due to the enhanced microstructural stability in the vicinity of the grain boundaries [52].

3.5.8. Summary—Creep Strength

In summary, Figure 3.20 graphically illustrates the creep strength of the conventional and new magnesium alloys in comparison with the aluminum (A380) alloy. The graph evidently revealed that the new AX (Mg–Al–Ca) and AXJ (Mg–Al–Ca–Sr) alloys exhibit significantly higher creep strength than their other (AZ, AS, and AE) magnesium-based counterparts. The creep strength of new magnesium alloys remained only marginally inferior (~10–20% less) to that of Al (A380) alloy which is the leading material used in powertrain systems.

3.6. MAGNESIUM-BASED BULK METALLIC GLASSES

In recent years, bulk metallic glasses (BMGs) based on magnesium alloy systems are gaining attention with the increasing demand for light and strong materials which are

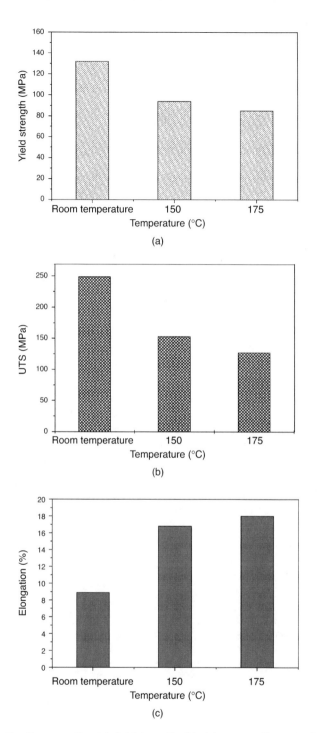

Figure 3.18. Tensile properties: (a) yield strength, (b) ultimate tensile strength, and (c) elongation of AS41 (Mg–Al–Si) alloy at room temperature, 150°C and 175°C (data extracted from [33]).

TABLE 3.27. Mechanical properties of AS21 (Mg–Al–Si) alloy [25, 48].

Alloy	Compressive YS (MPa)		Tensile YS (MPa)		UTS (MPa)		Elongation (%)		Fatigue Strength (MPa)	Impact Strength (J)
	20°C	150°C	RT	150°C	RT	150°C	RT	150°C		
AS21	125	87	120	90	235	125	12	35	90	14

RT, room temperature.

capable of withstanding severe environmental conditions. In comparison with their crystalline counterparts that usually exhibit poor corrosion resistance and are brittle due to their hexagonal close-packed (hcp) crystal structure, magnesium-based BMGs possess improved corrosion resistance and higher strength [53, 54]. Improved corrosion resistance and higher strength of BMGs can be attributed to [53, 54]

(i) their intrinsic structural and chemical homogeneities, and
(ii) the absence of crystalline slip system.

Magnesium-based BMGs were reported to have about four times higher strength than AZ91 (a typical high specific strength magnesium-based crystalline alloy) [55–58]. These improved properties coupled with their low cost and ability to be recycled, make magnesium-based BMGs highly sought after for engineering applications.

However, all the monolithic magnesium-based BMGs reported in existing literature [58–66] suffer from low ductility issue. Under compression loading, such materials always fail in the elastic regime with no observable plasticity (see Table 3.29).

In order to enhance the ductility of magnesium-based BMGs, successful attempts have been made to synthesize amorphous matrix composites. Second phases such as refractory ceramic or metallic particles are homogeneously dispersed in the BMG matrix in an attempt to improve the ductility of BMGs. There are two synthesis approaches:

(i) *In situ* method [67–69]
(ii) *Ex situ* method [70]

TABLE 3.28. Corrosion rate and total creep extension (ε_t) of AS41, AS21 (Mg–Al–Si), AZ91D, and Al alloy (A380) [33, 48].

Alloy	Corrosion Rate (mg cm^{-2} per day)	ε_t, 35 MPa/200 h		ε_t, 50 MPa/150°C	
		150°C	175°C	200 h	500 h
AS41	0.16	0.13	0.50	0.45	0.74
AS21	0.34	—	—	—	—
AZ91D	0.10	1.21	1.84	2.70	6.35
A380	0.34	0.18	0.15	0.08	0.10

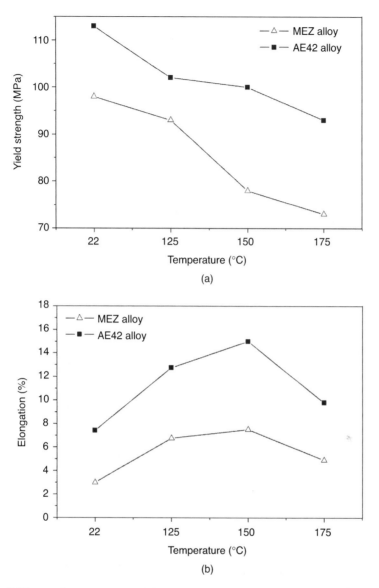

Figure 3.19. Tensile properties: (a) yield strength and (b) elongation of MEZ (Mg–2.5RE–0.35Zn–0.3Mn) and AE42 alloys at different testing temperatures (data extracted from [52]).

To date, a variety of metallic and ceramic particles, namely, ZrO_2 [71], WC [72], SiC [73], TiB_2 [70], Nb [64,74], Be [69], Fe [67], Ti [75], are used to develop Mg-based bulk metallic glass matrix composites (BMGMC).

Ma et al. [67] successfully synthesized *in situ* Mg-based BMGMC consisting of a Mg-based metallic glass matrix ($Mg_{65}Cu_{7.5}Ni_{7.5}Zn_5Ag_5Y_{10}$ alloy) with homogeneously

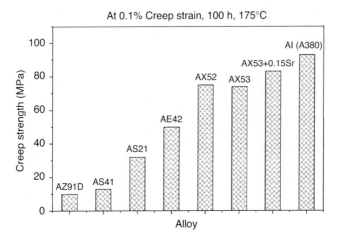

Figure 3.20. Creep strength of magnesium and aluminum die casting alloys to produce 0.1% creep strain in 100 h, at 175°C (data extracted from [28, 32]).

distributed Fe-rich particles. It was observed that the Fe addition had no adverse effect on the glass forming ability of the Mg-based matrix and its presence resulted in high compressive strength (~1 GPa) with observable yielding.

In another study by Xu et al. [70], $Mg_{65}Cu_{7.5}Ni_{7.5}Zn_5Ag_5Y_{10}$ alloy was reinforced with TiB_2 particles. The synthesized Mg-based BMG composite exhibited a plastic strain to failure of 2–3% and a high compressive strength of 1.3 GPa. The TiB_2 particles act as obstacles to the propagation of microcracks and run-away shear bands. The large elastic mismatch between the BMG matrix and TiB_2 particles also improve the plasticity through the generation of multiple shear bands.

In a study by Kinaka et al. [75], immiscible Ti powders were incorporated into $Mg_{65}Cu_{25}Gd_{10}$ BMG to achieve high compressive strength of ~900 MPa, as well as significant plastic deformation of ~40%. This achievement of simultaneous improvement

TABLE 3.29. Compression test results of monolithic magnesium-based BMGs [58–66].

Alloy (at.%)	Fracture Stress (MPa)	Plastic Strain (%)	Reference
$Mg_{65}Cu_{15}Ag_5Pd_5Y_{10}$	770	0	[59]
$Mg_{65}Cu_{20}Ag_{10}Y_2Gd_8$	956	0.3	[60]
$Mg_{65}Cu_{20}Ag_5Gd_{10}$	909	0.5	[61]
$Mg_{65}Cu_{7.5}Ni_{7.5}Zn_5Ag_5Y_{10}$	832	0	[62]
$Mg_{75}Cu_5Ni_{10}Gd_{10}$	874	0.2	[58]
$Mg_{65}Cu_{25}Gd_{10}$	834	0	[63]
$Mg_{65}Ni_5Cu_{20}Gd_{10}$	904	0.15	[63]
$Mg_{58.5}Cu_{30.5}Y_{11}$	1022	0.35	[64]
$Mg_{61}Cu_{28}Gd_{11}$	1075	0.4	[65]
$Mg_{65}Cu_{10}Ag_5Gd_{10}Ni_{10}$	1005	0	[66]

in strength and ductility opens the door to potential applications of such Mg-based BMGMCs, whereby high specific strength and high ductility are both essential.

REFERENCES

1. M. M. Avedesian and H. Baker (eds) (1999) *ASM Specialty Handbook—Magnesium and Magnesium Alloys*. Materials Park, OH: ASM International.
2. C. D. Yim, B. S. You, R. S. Jang, and S. G. Lim (2006) Effects of melt temperature and mold preheating temperature on the fluidity of Ca containing AZ31 alloys. *Journal of Materials Science*, **41**(8), 2347–2350.
3. L. Han, H. Hu, D. O. Northwood, and N. Li (2008) Microstructure and nanoscale mechanical behavior of Mg-Al and Mg-Al-Ca alloys. *Materials Science and Engineering A*, **473**(1–2), 16–27.
4. R. K. Mishra, A. K. Gupta, P. R. Rao, A. K. Sachdev, A. M. Kumar, and A. A. Luo (2008) Influence of cerium on the texture and ductility of magnesium extrusions. *Scripta Materialia*, **59**(5), 562–565.
5. A. A. Nayeb-Hashemi and J. B. Clark (eds) (1988) *Phase diagrams of binary magnesium alloys*. Metals Park, OH: ASM International.
6. S. F. Hassan and M. Gupta (2002) Development of a novel magnesium-copper based composite with improved mechanical properties. *Materials Research Bulletin*, **37**, 337–389.
7. S. F. Hassan and M. Gupta (2003) Development of high strength magnesium-copper based hybrid composites with enhanced tensile properties. *Materials Science and Technology*, **19**, 253–259.
8. W. L. E. Wong and M. Gupta (2007) Development of Mg/Cu nanocomposites suing microwave assisted rapid sintering. *Composites Science and Technology*, **67**, 1541–1552.
9. W. L. E. Wong and M. Gupta (2005) Enhancing thermal stability, modulus and ductility of magnesium using molybdenum as reinforcement. *Advanced Engineering Materials*, **7**(4), 250–256.
10. S. F. Hassan and M. Gupta (2002) Development of a novel magnesium/nickel composite with improved mechanical properties. *Journal of Alloys and Compounds*, **335**, L10–L15.
11. P. Lyon, I. Syed, and S. Heaney. Elektron 21—an aerospace magnesium alloy for sand cast and investment cast applications. *Advanced Engineering Materials*, **9**(9), 793–798.
12. B. Powell, A. Luo, V. Rezhets, J. Bommarito, and B. Tiwari (2001) Development of creep-resistant magnesium alloys for powertrain applications: Part 1. SAE Technical Paper 2001–01-0422. Warrendale, PA: Society of Automotive Engineers.
13. S. F. Hassan and M. Gupta (2002) Development of ductile magnesium composite materials using titanium as reinforcement. *Journal of Alloys and Compounds*, **345**, 246–251.

14. P. Lyon, T. Wilks, and I. Syed (2005) The influence of alloying elements and heat treatment upon properties of Elektron 21 (EV31A) alloy. In N. R. Neelameggham, H. I. Kaplan, and B. R. Powell (eds) *TMS Magnesium Technology*. Warrendale, PA, pp. 303–308.

15. ASM Handbook Volume 15: Casting, p. 798.

16. K. U. Kainer (ed.) (2003) *Magnesium–Alloys and Technologies*. Weinheim, Cambridge: Wiley-VCH.

17. ASTM B296 – 03 (2008): Standard Practice for Temper Designations of Magnesium Alloys, Cast and Wrought.

18. R. E. Brown (2006) In M. Kutz (ed.) *Mechanical Engineers' Handbook: Materials and Mechanical Design*, Vol. **1**, 3rd edition. John Wiley & Sons, pp. 278–286.

19. H. E. Friedrich and B. L. Mordike (eds) (2006) *Magnesium Technology–Metallurgy, Design Data, Applications*. Springer, p. 93.

20. F. S. Pan, A. T. Tang, S. Y. Long, and M. B. Yang (2007) Development and application of wrought magnesium alloys in China. In K. U. Kainer (ed.) *Magnesium, Proceedings of the 7th International Conference on Magnesium Alloys and Their Applications*, Pp. 297–304.

21. A. Stalmann, W. Sebastian, H. Friedrich, S. Schumann, and K. Dröder (2001) Properties and processing of magnesium wrought products for automotive applications. *Advanced Engineering Materials*, **3**(12), 969–974.

22. Magnesium Elektron Website: http://www.magnesium-elektron.com/ (last accessed on June 10, 2008).

23. A. A. Luo (2000) Materials comparison and potential applications of magnesium in automobiles. In H. I. Kaplan, J. N. Hryn, and B. B. Clow (eds) *Magnesium Technology 2000*. Warrendale, PA: TMS, pp. 89–98.

24. Magnesium Die Casting Handbook (1998) NADACA.

25. A. A. Luo (2004) Recent magnesium alloy development for elevated temperature applications. *International Materials Reviews*, **49**(1), 13–30.

26. M. O. Pekguleryuz and A. A. Kaya (2003) Creep resistant magnesium alloys for powertrain applications. *Advanced Engineering Materials*, **5**(12), 866–878.

27. C. Suma (1991) Creep of diecast magnesium alloys AZ91D and AM60B. SAE Technical Paper 910416. Warrendale, PA: Society of Automotive Engineers.

28. W. E. Mercer II (1990) Magnesium die cast alloys for elevated temperature applications. SAE Technical Paper 900788. Warrendale, PA: Society of Automotive Engineers.

29. P. Lyon, J. F. King, and K. Nuttall (1996) A new magnesium HPDC alloy for elevated temperature use. In G. W. Lorimer (ed.) *Proceedings of 3rd International Magnesium Conference*. London: The Institute of Materials, pp. 99–108.

30. M. O. Pekguleryuz and A. Luo. Creep resistant magnesium alloys for die casting. PCT (Patent Cooperative Treaty) application WO 96/25529, filed 14 Feb 1996, published 22 Aug 1996.

31. A. Luo and T. Shinoda (1998) A new magnesium alloy for automotive powertrain applications. SAE Technical Paper 980086. Warrendale, PA: Society of Automotive Engineers.

32. A. Luo, M. Balogh, and B. Powell (2001) Development of creep-resistant magnesium alloys for powertrain applications: Part 2 of 2. SAE Technical Paper 2001–01–0423. Warrendale, PA: Society of Automotive Engineers.

33. M. O. Pekguleryuz and E. Baril (2001) Development of creep resistant Mg-Al-Sr alloys. In J. N. Hryn (ed.) *Magnesium Technology 2001*. Warrendale, PA: The Minerals, Metals and Materials Society, pp. 119–125.

34. K. Makino, T. Kawata, K. Kanemitsu, and K. Watanabe (1994). Magnesium Alloy. UK patent Application, GB 2 296256 A, filed 22 Dec 1994, published 26 June 1996.

35. K. Sakamoto, Y. Yamamoto, N. Sakate, and S. Hirabara (1997). Heat-resistant magnesium alloy member. European Patent Application, EP 0 799 901 A1, filed 4 April 1997, published 10 Aug 1997.

36. T. Baba, K. Honma, and M. Ichikawa (1997). Heat-resistant Magnesium Alloy. European Patent Application, EP 0 791 662 A1, filed 2 Feb 1997, published 27 Aug 1997.

37. S. Park, J. J. Kim, D. H. Kim, C. S. Shin, and N. J. Kim (1997). Magnesium Alloy for a High Pressure Casting and Process for Preparation Thereof. PCT (Patent Cooperative Treaty) application WO 97/40201, filed 25 April 1997, published 30 Oct 1997.

38. D. Argo, M. O. Pekguleryuz, P. Labelle, M. Direks, T. Sparks, and T. Waltemate (2001) Die castability and properties of Mg-Al-Sr based alloys. In J. N. Hryn (ed.) *Magnesium Technology 2001*. Warrendale, PA: TMS, pp. 131–136.

39. B. Bronfin, E. Aghion, F. von Buch, S. Schumann, and M. Katsir (2001) Die casting magnesium alloys for elevated temperature applications. In J. N. Hryn (ed.), *Magnesium Technology 2001*. Warrendale, PA: TMS, pp. 127–130.

40. K. Pettersen, P. Bakke, J. I. Skar, C. Tian, and D. Albright (2001) *Transactions of 21st International Die Casting Congress*. Rosemont, IL: North American Die Casting Association (NADCA), pp. 181–186.

41. B. R. Powell, V. Rezhets, M. P. Balogh, and R. A. Waldo (2002) Microstructure and creep behavior in AE42 magnesium die casting alloy. *JOM*, **54**(8), 34–38.

42. A. A. Luo and M. O. Pekguleryuz (1994) Cast magnesium alloys for elevated temperature applications. *Journal of Materials Science*, **29**, 5259–5271.

43. R. Ninomiya, T. Ojiro, and K. Kubota (1995) Improved heat resistance of Mg-Al alloys by the Ca addition. *Acta Metallurgica et Materialia*, **43**(2), 669–674.

44. M. O. Pekguleryuz and A. Luo. Creep Resistant Magnesium Alloys for Diecasting. PCT (Patent Cooperative Treaty) application No. PCT/CA96/00091, published 22 Aug 1996.

45. M. O. Pekguleryuz and J. Renaud (2000) Creep resistance in Mg-Al-Ca alloys. In H. Kaplan, J. Hryn, and B. Clow (eds), *Magnesium Technology 2000*, TMS, pp. 279–284.

46. B. R. Powell, V. Rezhets, A. A. Luo, J. J. Bommarito, and B. L. Tiwari. Creep-resistant magnesium alloy die casting. US Patent 6,264,763, 24 July 2001.

47. B. Bronfin, E. Aghion, S. Schuman, P. Bohling, and K. U. Kainer. Magnesium alloy for high temperature applications. US Patent 6139651, 31 Oct 2000.

48. *MRI 153–New Magnesium Alloy for High Temperature Applications* (2001) Beersheva, Israel: Dead Sea Magnesium.

49. M. Pekguleryuz, P. Labelle, E. Baril, and D. Argo (2003) Magnesium diecasting alloy AJ62x with superior creep resistance and castability. *Magnesium Technology 2003*. San Diego: TMS, pp. 201–207.

50. S. Beer, G. Frommeyer, and E. Schmid (1992) Development of Mg-Mg$_2$Si light weight alloys. In B. L. Mordike and F. Hehmann (eds) *Magnesium Alloys and Their Applications*. Oberursel: DGM Informationsgesellschaft, pp. 317–324.

51. J. F. King (1998) Development of magnesium diecasting alloys. In B. L. Mordike and K. U. Kainer (eds) *Magnesium Alloys and Their Applications*. Frankfurt: Wekstoff-Informationsgesellschaft, pp. 37–47.

52. I. P. Moreno, T. K. Nandy, J. W. Jones, J. E. Allison, and T. M. Pollock (2002) Microstructure and creep behavior of a die cast magnesium-rare earth alloy. In H. I. Kaplan (ed.), *Magnesium Technology 2002*. Warrendale, PA: TMS, pp. 111–116.

53. A. Inoue (2000) Stabilization of metallic supercooled liquid and bulk amorphous alloys. *Acta Materialia*, **48**, 279–285.

54. W. L. Johnson (1999) Bulk glass-forming metallic alloys: science and technology. *MRS Bulletin*, **24**, 42–56.

55. E. S. Park, H. G. Kang, W. T. Kim, and D. H. Kim (2001) The effect of Ag addition on the glass-forming ability of Mg-Cu-Y metallic glass alloys. *Journal of Non-crystalline Solids* **279**, 154–160.

56. H. Men, Z. Q. Hu, and J. Xu (2002) Bulk metallic glass formation in the Mg-Cu-Zn-Y system. *Scripta Materialia*, **46**(10), 699–703.

57. H. Ma, Q, Zheng, J. Xu, Y. Li, and E. Ma (2005) Doubling the critical size for bulk metallic glass formation in the Mg-Cu-Y ternary system. *Journal of Materials Research*, **20**(9), 2252–2255.

58. G. Yuan, C. Qin, and A. Inoue (2005) Mg-based bulk glassy alloys with high strength above 900 MPa and plastic strain. *Journal of Materials Research*, **20**, 394–400.

59. K. Amiya and A. Inoue (2000) Thermal stability and mechanical properties of Mg-Y-Cu-M (M = Ag, Pd) bulk amorphous alloys. *Materials Transactions*, **41**, 1460–1462.

60. E. S. Park, W. T. Kim, and D. H. Kim (2004) Bulk glass formation in Mg-Cu-Ag-Y-Gd alloy. *Materials Transactions*, **45**, 2474–2477.

61. E. S. Park, J. Y. Lee, and D. H. Kim (2005) Effect of Ag addition on the improvement of glass-forming ability and plasticity of Mg-Cu-Gd bulk metallic glass. *Journal of Materials Research*, **20**(9), 2379–2385.

62. E. S. Park and D. H. Kim (2005) Formation of Mg-Cu-Ni-Ag-Zn-Y-Gd bulk glassy alloy by casting into cone-shaped copper mold in air atmosphere. *Journal of Materials Research*, **20**(6), 1465–1469.

63. G. Yuan and A. Inoue (2005) The effect of Ni substitution on the glass-forming ability and mechanical properties of Mg-Cu-Gd metallic glass alloys. *Journal of Alloys and Compounds*, **387**, 134–138.

64. Q. Zheng, H. Ma, E. Ma, and J. Xu (2006) Mg-Cu-(Y,Nd) pseudo-ternary bulk metallic glasses: The effects of Nd on glass-forming ability and plasticity. *Scripta Materialia*, **55**(6), 541–544.

65. Q. Zheng, S. Cheng, J. H. Strader, E. Ma, and J. Xu (2007) Critical size and strength of the best bulk metallic glass former in the Mg-Cu-Gd ternary system. *Scripta Materaila*, **56**(2), 161–164.

66. D. G. Pan, W. Y. Liu, H. F. Zhang, A. M. Wang, and Z. Q. Hu (2007) Mg-Cu-Ag-Gd-Ni bulk metallic glass with high mechanical strength. *Journal of Alloys and Compounds*, **438**, 142–144.

67. H. Ma, J. Xu, and E. Ma (2003) Mg-based bulk metallic glass composites with plasticity and high strength. *Applied Physics Letters*, **83**(14), 2793–2795.

68. X. Hui, W. Dong, G. L. Chen, and K. F. Yao (2007) Formation, microstructure and properties of long-period order structure reinforced Mg-based bulk metallic glass composites. *Acta Materialia*, **55**(3), 907–920.

69. Z. G. Li, X. Hui, C. M. Zhang, M. L. Wang, and G. L. Chen (2007) Strengthening and toughening of Mg-Cu-Y,Gd) bulk metallic glasses by minor addition of Be. *Materials Letters*, **61**(28), 5018–5021.

70. Y. K. Xu, H. Ma, J. Xu, and E. Ma (2005) Mg-based bulk metallic glass composites with plasticity and gigapascal strength. *Acta Materialia*, **53**(6), 1857–1866.

71. J. S. C. Jang, L. J. Chang, J. H. Young, J. C. Huang, and C. Y. A. Tsao (2006) Synthesis and characterization of the Mg-based amorphous/nano ZrO_2 composite alloy. *Intermetallics*, **14**(8–9), 945–950.

72. P. Y. Lee, C. Lo, J. S. C. Jiang, and J. C. Huang (2006) Mg-Y-Cu bulk nanocrystalline matrix composites containing WC particles. *Key Engineering Materials*, **313**, 25–30.

73. J. Li, L. Wang, H. Zhang, Z. Hu, and H. Cai (2007) Synthesis and characterization of particulate reinforced Mg-based bulk metallic glass composites. *Materials Letters*, **61**(11–12), 2217–2221.

74. D. G. Pan, H. F. Zhang, A. M. Wang, and Z. Q. Hu (2006) Enhanced plasticity in Mg-based bulk metallic glass composite reinforced with ductile Nb particles. *Applied Physics Letters*, **89**(26), 261904.

75. M. Kinaka, H. Kato, M. Hasegawa, and A. Inoue (2008) High specific strength Mg-based bulk metallic glass matrix composite highly ductilized by Ti dispersoid. *Materials Science and Engineering A*, **494**, 299–303.

4

FUNDAMENTALS OF METAL MATRIX COMPOSITES

This chapter presents the fundamentals related to metal matrix composites. Metal matrix composites represent the unified combination of a metallic matrix and reinforcement that can be either discontinuous or continuous in nature. The effects of the choice of metallic matrices and type of reinforcements on the overall performance of the resultant composite are highlighted. The matrix–reinforcement interfacial influence is also presented. Lastly, theoretical models for prediction of physical, electrical, thermal, and mechanical properties are discussed.

4.1. INTRODUCTION

Composites can be defined as materials that contain two or more distinct materials as a unified combination. Composite materials have the ability to combine the properties of reinforcing phase with that of the matrix phase such that the resultant properties of the composite materials are better than the properties of their monolithic counterparts. Their properties can be tailored, depending on end application, through the judicious selection of reinforcement phase, matrix phase, and processing technique.

Magnesium, Magnesium Alloys, & Magnesium Composites, by Manoj Gupta and Nai Mui Ling, Sharon
© 2010 John Wiley & Sons, Inc.

As a result, composite materials have the capability to serve a wide spectrum of applications.

Composite materials can be classified into three broad categories as follows:

(i) Polymer matrix composites (PMCs)
(ii) Ceramic matrix composites (CMCs)
(iii) Metal matrix composites (MMCs)

PMCs have extensive range of applications, from uses in aerospace structures to sports equipment (such as golf shafts and tennis racket handles). CMCs find applications especially in high-temperature components. MMCs are increasingly sought for a wide range of applications in the aerospace, automotive, and structural fields because of their several attributes. MMCs offer higher ductility than CMCs and better environmental stability than PMCs. MMCs can also be recycled, and hence, are more sustainable.

In comparison to monolithic metallic materials, MMCs offer the following advantages:

- Higher strength to weight ratio, which results in significant weight savings.
- Higher specific modulus.
- Enhanced elevated temperature stability in terms of better creep resistance.
- Improved dimensional stability (e.g., $CTE_{Mg} = 28.73 \times 10^{-6} \, °C^{-1}$, whereas $CTE_{Mg/2vol.\%Y_2O_3} = 26.82 \times 10^{-6} \, °C^{-1}$ [1]).
- Comparable or improved fatigue characteristics.
- Enhanced damping characteristics.
- Enhanced abrasion and wear resistance.

4.1.1. Factors Affecting Properties of MMCs

The properties of MMCs can be tailored to suit end applications. A number of factors decide the end properties and their optimization is a must to realize the desired combination of properties. Some of the key factors are listed below:

(i) Properties of matrix phase
(ii) Type of reinforcing phase
(iii) Amount of reinforcing phase
(iv) Size of reinforcing phase
(v) Morphology of reinforcing phase
(vi) Orientation of reinforcing phase
(vii) Distribution of reinforcing phase
(viii) Nature of reinforcement–matrix interface

(ix) Heat treatment procedure

(x) Type of processing technique

4.2. MATERIALS

MMCs consist of

(i) a matrix material that is either a metal or a metallic alloy and

(ii) a reinforcement material that can be metallic, ceramic, refractory metal, intermetallic, or semiconductor.

Types of MMCs can be grouped according to the form of the reinforcements, as presented
in Figure 4.1:

(i) Particulates

(ii) Short fibers or whiskers

(iii) Continuous fiber

(iv) Laminates or sheets

(v) Interconnected reinforcement

(vi) Singular metal core reinforcement

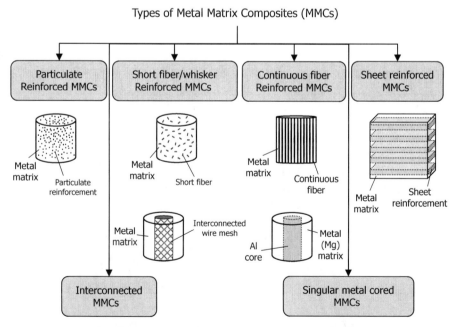

Figure 4.1. Different types of metal matrix composites.

TABLE 4.1. List of typical metallic matrices [2–5].

Metallic Matrices
Aluminum and its alloys
Bronze
Copper
Cobalt
Lead
Magnesium and its alloys
Nickel
Niobium
Silver
Sn-based Solders
Superalloys (Ni-based and Fe-based)
Titanium alloys
Intermetallics (Nickel aluminides and Titanium aluminides)

Particulate or discontinuously reinforced MMCs assumed significant interest because of their advantages over other types of MMCs, which include

(i) the ability to exhibit isotropic properties,

(ii) the ability to be synthesized/fabricated using conventional metallurgical processing,

(iii) the ability to be machined using conventional methods, and

(iv) low cost.

4.2.1. Matrix

Matrix phase in a composite material is a continuous phase that surrounds and binds the reinforcing phase into a three-dimensional geometrical form of required dimensions. It also functions to take up the external load and to transfer the load to the reinforcing phase. Table 4.1 lists some of the typical metallic matrices. Aluminum and its alloys are one of the most commonly used matrix materials because of the increasing demand for lightweight and high-strength materials. In recent years, magnesium and titanium alloys have also seen a surge in usage as matrix materials for composites because of their lightweight characteristics. However, owing to the reactivity of these metals, the synthesis of such composites is challenging. In addition to the lightweight metals, other metals/alloys such as bronze, cobalt, copper, lead, nickel, niobium, silver, superalloys (Ni-based and Fe-based) [2], and Sn-based solders [3–5] are also utilized as matrix materials for a variety of specific end applications.

4.2.2. Reinforcements

A wide range of reinforcements has been successfully incorporated into metallic matrices to form MMCs. Table 4.2 lists some of the commonly used reinforcements and their

TABLE 4.2. Properties of some commonly used reinforcements [7–14].

Reinforcement	Density (× 10^{-3} kg m^{-3})	CTE (×10^{-6}°C^{-1})	Elastic Modulus (GPa)		Strength (MPa)		Reference
			(GPa)	(°C)	(MPa)	(°C)	
Al$_2$O$_3$	3.98	7.92	379	1090	221	1090	[7]
AlN	3.26	4.84	310	1090	2069	24	[7]
BeO	3.01	7.38	190	1090	24	1090	[7]
B$_4$C	2.52	6.08	448	24	2759	24	[7]
C	2.18	−1.44	690	—	—	—	[7]
CeO$_2$	7.13	12.42	185	24	589	24	[7]
CNT	1.33–1.40	—	~1000	—	~30,000	—	[8]
Cu	8.94	18.3	129.8	—	—	—	[9]
HfC	12.20	6.66	317	24	—	—	[7]
MgO	3.58	11.61	317	1090	41	1090	[7]
Mo	10.22	4.8 K-1	324.8	—	—	—	[8]
MoSi$_2$	6.31	8.91	276	1260	276	1090	[7]
Mo$_2$C	8.90	5.81	228	24	—	—	[7]
Ni	8.90	13.9	207	—	—	—	[10,11]
NbC	7.60	6.84	338	24	—	—	[7]
Si	2.33	3.06	112	—	—	—	[7]
SiC	3.21	5.40	324	1090	—	—	[7]
Si$_3$N$_4$	3.18	1.44	207	—	—	—	[7]
SiO$_2$	2.66	<1.08	73	—	—	—	[7]
SnO$_2$	6.95	3.76	233	—	—	—	[3,12]
TaC	13.90	6.46	366	24	—	—	[7]
TaSi$_2$	—	10.80	338	1260	—	—	[7]
ThO$_2$	9.86	9.54	200	1090	193	1090	[7]
Ti	4.51	9.1	102.2	—	586[a]	—	[10,11]
TiB$_2$	4.50	8.28	414	1090	—	—	[7]
TiC	4.93	7.60	269	24	55	1090	[7]
UO$_2$	10.96	9.54	172	1090	—	—	[7]
VC	5.77	7.16	434	24	—	—	[7]
WC	15.63	5.09	669	24	—	—	[7]
WSi$_2$	9.40	9.00	248	1090	—	—	[7]
Y$_2$O$_3$	5.03	—	171.5	—	—	—	[13,14]
ZrB$_2$	6.09	8.28	503	24	—	—	[7]
ZrC	6.73	6.66	359	24	90	1090	[7]
ZrO$_2$	5.89	12.01	132	1090	83	1090	[7]

CTE, coefficient of thermal expansion.
[a]Value of yield strength.

properties. In order to ensure the right choice of reinforcement, the following factors should be taken into consideration during the selection [6, 7]:

  (i) Size and morphology of reinforcement
 (ii) Physical properties of reinforcement:
   - Density
   - Melting temperature (T_m)
   - Coefficient of thermal expansion (CTE)
   - Thermal conductivity
   - Electrical conductivity
   - Wettability with matrix material (compatibility with matrix material)
(iii) Mechanical properties of reinforcement:
   - Elastic modulus (E)
   - Strength
   - Hardness
 (iv) Cost of reinforcement

These factors will, in turn, influence the end application of the composite and the synthesis method of the metal matrix composite. For use as structural materials, the performance of the MMCs is dependent on the density, elastic modulus, and tensile strength of the reinforcements. For thermal management applications, the thermal conductivity and coefficient of thermal expansion properties of the reinforcing phase are of important consideration.

As earlier presented in Chapter 2, there are several ways to fabricate the MMCs. These processing techniques can be broadly classified into the powder metallurgy and the liquid metallurgy methods. For composites synthesized using the powder metallurgy method, the matrix powder is blended with the reinforcement powder. In order to avoid agglomeration and to achieve a homogenous composite mixture, it is essential to judiciously select the size of the matrix and reinforcement powders [6]. Moreover, when incorporating the ceramic reinforcements into the metallic matrix, their brittle characteristic often results in particle fracture during the powder metallurgy process. It has been reported that the right selection of particle aspect ratio and particle size will minimize this issue [6].

For composites synthesized using the liquid metallurgy method, the reinforcing phases are in constant contact with the molten metallic matrix over a considerable duration and under high processing temperature. These can lead to the reaction between the two phases [6]. This reaction can be detrimental to the overall performance of the MMC; thus, the choice of reinforcement is important to match the choice of matrix material, so as to realize the full potential of composite material. In the case of aluminum alloys, the incorporation of SiC (which is thermodynamically unstable in most molten Al alloys) will result in the formation of Al_4C_3. When Al_2O_3 reinforcements are introduced into Mg-free Al alloys, Al_2O_3 is thermodynamically stable. On the contrary, SiC is stable in molten magnesium alloys, whereas Al_2O_3 reacts in Mg alloys to form Al_2MgO_4 spinel [6].

The size and density of the reinforcing phase are also important factors to consider as large-size reinforcements and those with heavier density than that of the matrix have the tendency to settle [6]. On the other hand, smaller-size reinforcements and those with lighter density than that of the matrix have the tendency to float, according to the Stokes' law presented in equation (4.1) [15]. In both cases, a segregated casting is resulted which is undesirable except for the case of functionally gradient composite materials [16, 17].

According to Stokes' law, the drag forces (F_d) on a spherical-shaped particle can be expressed as follows [15]:

$$F_d = 6\pi\mu R V \qquad (4.1)$$

where V is the velocity of the reinforcing particles (which is downward if $\rho_r > \rho_m$ and upward if $\rho_r < \rho_m$), μ is the viscosity of the molten metal, and R is the radius of the particle.

If the reinforcing particles are falling in the viscous melt by their own weight because of gravitational forces, then a terminal velocity, which is also known as the settling velocity, is reached when the drag forces together with the buoyant forces balanced out the gravitational forces:

$$\frac{4}{3}\pi g R^3 (\rho_r - \rho_m) = 6\pi\mu R V \qquad (4.2)$$

where ρ_r and ρ_m represent the density of the reinforcing particle and the matrix, respectively, whereas g is the gravitational acceleration. Hence, the resultant settling velocity of the particle is defined as

$$V = \frac{2}{9}\left[\frac{(\rho_r - \rho_m)}{\mu}\right](g R^2) \qquad (4.3)$$

Moreover, larger reinforcements are easier to incorporate into the molten melt as compared to their nanosize counterparts. The latter are prone to agglomeration among themselves [18], and hence, effort must be made to ensure the uniform distribution of reinforcements within the matrix material during the synthesis of the composite material. Furthermore, the amount of reinforcements introduced into the matrix material will affect the viscosity of the melt. With increasing amount of reinforcements, the viscosity of the melt increases accordingly, making the synthesis process more challenging.

For composite materials, independent of the synthesis methodology, a certain threshold exists, whereby too much reinforcement will degrade the properties of the composite material. Thus, judicious selection of the reinforcement and its related factors are crucial.

4.3. INTERFACE BETWEEN MATRIX AND REINFORCEMENT

The interface formed between the metallic matrix and the reinforcement plays a crucial role in the MMCs. In most cases, the reinforcement and matrix are not in thermodynamic equilibrium and there exists a thermodynamic driving force for an interfacial reaction to take place to lower the overall system's energy. The characteristics of the interface influence the crack resistance and load transfer during deformation. These will, in turn, affect the properties of the MMCs. If the interfacial bond is weak, the interface will fail before any effective stress transfer occurs and no strengthening is achieved. The composite may be weaker than the unreinforced material as the effective area supporting the load is now reduced.

In order to enhance the bond strength of the interface, it is essential that

  (i) wetting between the matrix material and the reinforcement is promoted,
  (ii) the formation of oxide is minimized, and
  (iii) chemical interactions are controlled.

The two types of interfacial bonding between the reinforcement and the matrix in a metal matrix composite can be broadly grouped into

  (i) mechanical bonding (interlocking) and
  (ii) chemical bonding.

As defined by the Young's equation [19], the wettability of a solid by a liquid is determined by the angle of contact which the liquid makes on the solid as shown in Figure 4.2. This contact angle (θ) is defined from the following Young's equation.

$$\gamma_{sg} - \gamma_{sl} = \gamma_{lg} \cos \theta \tag{4.4}$$

$$\theta = \cos^{-1} \left(\frac{\gamma_{sg} - \gamma_{sl}}{\gamma_{lg}} \right) \tag{4.5}$$

where γ_{sl}, γ_{sg}, and γ_{lg} are the interfacial energies between solid and liquid, solid and gas, and liquid and gas phases, respectively.

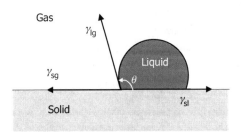

Figure 4.2. Contact angle (θ) formed between the liquid, solid, and gas phases.

In the case when $\theta < 90°$ ($\gamma_{sg} > \gamma_{sl}$), the solid is wetted by the liquid matrix, which implies that wetting took place. When $\theta = 0°$, there is perfect wetting, and when $\theta = 180°$, there is no wetting.

The effect of wettability on the interface adhesion can be determined by the work of adhesion, W_{ad} [20]. Good wetting is denoted by a high work of adhesion, whereas poor wetting is denoted by a low work of adhesion:

$$W_{ad} = \gamma_{lg}(1 + \cos\theta) \qquad (4.6)$$

Wetting between the ceramic reinforcement and the metallic matrix is usually difficult due to the high surface tension of the molten metals. Mortensen et al. [21] reported that wetting is generally poor between reinforcements (C, SiC, Al_2O_3, and B_4C) with aluminum and aluminum alloys at temperatures below 950°C. In order to improve wetting in metal matrix composite systems, the contact angle should be reduced by

(i) decreasing the solid–liquid interfacial energy (γ_{sl}),
(ii) decreasing the liquid metal's surface tension, and
(iii) increasing the solid's surface energy.

These can be achieved by

(i) heat treating the ceramic particulates,
(ii) coating the ceramic particulates with a metallic layer (for instance, in the Al–Al_2O_3 composite, the Al_2O_3 reinforcements are coated with Ni and Ti–Ni to promote wetting [22]), and
(iii) introducing reactive materials as alloying elements into the metallic matrix [7].

4.3.1. Tailoring the Interface for Enhanced Performance

From a technological standpoint of property–performance relationship, the interface between the matrix and the reinforcing phase is of primary importance. Processing of MMCs sometimes allows tailoring of the interface between the matrix and the reinforcement in order to meet the property–performance requirements. The strong interface in composites is usually associated with the development of a strong chemical bond between the metallic matrix and the reinforcement [23–25]. The extent of the matrix–reinforcement interaction, however, needs to be strictly regulated as excessive chemical activity at the interface can lead to the degradation of reinforcement strength and can be detrimental to the composite's strength. Moreover, the extensive formation of a brittle interfacial reaction product (such as Al_4C_3 in the case of Al/SiC composite) is also undesirable, as it may act as crack initiation site and a path for crack propagation in the composite. The reacted interface may also increase the susceptibility of the composite to environmental attack. On the other hand, a limited interfacial reaction that results in the formation of a thin layer of the interfacial reaction product may be compatible with the preservation of the matrix–reinforcement integrity, hence aiding in realizing an improvement in the strength of the metal matrix composite [26, 27].

4.3.2. Methods of Interface Engineering

Several methods have been used to tailor the strength of the interfacial bond and the resulting performance of the metal matrix composite. They include [28, 29] the following:

    (i) Matrix alloying
   (ii) Modifying the reinforcement
  (iii) Modifying the temperature cycle of the composite synthesis process
  (iv) Heat treating the processed composite

In the case of reinforcement modification, the reinforcements are subjected to a simple heat treatment to eliminate any deleterious absorbed species, or the reinforcements are coated with a material that is easily wetted by the metallic matrix (for instance, the SiO_2 coating on the SiC reinforcement formed when heat treating the reinforcement in air). These processes influence the formation of a strong interfacial bond by altering the interfacial chemistry to promote the wetting of the reinforcement with the molten metal.

4.3.3. Tailoring the Interface Through Choice of Processing Technique

The increased interfacial reactivity that accompanies the improved wetting of the reinforcement is one of the conflicting requirements, which must be taken into account in the processing of the MMCs [6, 7]. Although, a high chemical affinity between the metallic matrix and the reinforcement is desired so that the two phases can combine spontaneously, and hence, ease the composite processing. On the other hand, to avoid unwanted interfacial chemical reactions, low chemical affinity is preferred. Therefore, controlling the extent of the interfacial reaction during processing of the composite is of paramount importance. In composite systems with very high interfacial reactivity, a rapid solidification process (such as the spray atomization and codeposition) should be utilized so as to minimize the duration that the matrix and reinforcement are in contact at the processing temperature [28].

4.3.4. Interfacial Failure

Imperfections at the interface between the matrix and the reinforcement can facilitate the nucleation of cavities and cracks during the loading of the composite [30]. Plastic flow is concentrated near such regions and large tensile stresses are developed, promoting the growth of the cavities and cracks. When bonding at the interface is incomplete, the tensile loads placed on the matrix cannot be effectively transferred to the reinforcement. This leads to a decrease in the load carrying capacity of the composite and results in premature failure of the composite. Some of the possible causes of interfacial imperfections include [30] the following:

   (i) Precipitation reactions
  (ii) Interfacial reactions
 (iii) Solute segregation
 (iv) Clustering of reinforcements

The geometric imperfections of the reinforcement also play a crucial part in interfacial failure. Nutt et al. [31, 32] reported that void nucleation via interfacial decohesion at the fiber ends resulted in the premature failure of Al/SiC whisker composites. Voids typically nucleated at the sharp corners of the fiber ends, as they are sites of severe stress concentration.

4.4. THEORETICAL PREDICTION OF PROPERTIES

4.4.1. Density

The theoretical density value of the composite material can be predicted using the simple rule of mixture (ROM) model. The ROM model is a function of the volume fraction and the density of the respective matrix and reinforcing phases. The density of the composite, ρ_{comp}, can be determined by [33]

$$\rho_{comp} = \rho_m V_m + \rho_r V_r \tag{4.7}$$

and

$$V_m + V_r = 1 \tag{4.8}$$

where ρ_m and ρ_r represent the density of the matrix and the reinforcement, respectively, whereas V_m and V_r represent the volume fraction of the matrix and the reinforcement, respectively, incorporated in the composite material.

Although density of composites computed using ROM are fairly accurate, some deviations can be explained especially when some elements in the matrix react extensively with the reinforcement, leading to formation of reaction products.

4.4.2. Electrical Conductivity

4.4.2.1. Rayleigh–Maxwell Equation. The electrical conductivity for discontinuously reinforced composites, K_{comp}, containing a small volume fraction of approximately spherical reinforcement can be predicted by the Rayleigh–Maxwell equation as follows [33, 34]:

$$K_{\text{comp}} = K_{\text{m}} \left(\frac{1 + 2V_{\text{r}} \left[\dfrac{1 - K_{\text{m}}/K_{\text{r}}}{2K_{\text{m}}/K_{\text{r}} + 1} \right]}{1 - V_{\text{r}} \left[\dfrac{1 - K_{\text{m}}/K_{\text{r}}}{2K_{\text{m}}/K_{\text{r}} + 1} \right]} \right) \tag{4.9}$$

where K_{r} and K_{m} represent the electrical conductivity of the reinforcement and the metallic matrix, respectively. V_{r} represents the volume fraction of the reinforcement.

The Rayleigh–Maxwell equation has its limitation, as it does not take into account the microstructural changes brought about because of the presence of reinforcement in the metallic matrix.

4.4.2.2. Kerner's Model.
The model proposed by Kerner [35] takes into consideration the effect of the plastic zone, which is created because of the presence of reinforcement. According to Kerner's model, the conductivity of the composite samples, K_{comp}, can be predicted as follows:

$$K_{\text{comp}} = \frac{K_{\text{m}} V_{\text{m}} + K_{\text{r}} V_{\text{r}} L_{\text{r}} + K_{\text{pl}} V_{\text{pl}} L_{\text{pl}}}{V_{\text{m}} + V_{\text{r}} L_{\text{r}} + V_{\text{pl}} L_{\text{pl}} + V_{\text{p}} L_{\text{p}}} \tag{4.10}$$

where

$$L_{\text{r}} = \frac{9 K_{\text{m}} K_{\text{pl}}}{(K_{\text{r}} + 2K_{\text{pl}})(K_{\text{pl}} + 2K_{\text{m}}) + \left(\dfrac{2V_{\text{r}}}{V_{\text{r}} + V_{\text{pl}}} \right)(K_{\text{m}} - K_{\text{pl}})(K_{\text{pl}} - K_{\text{r}})} \tag{4.11}$$

$$L_{\text{pl}} = \frac{2}{3} L_{\text{r}} \quad \text{and} \tag{4.12}$$

$$L_{\text{p}} = \frac{3}{2} \text{ since inside the porosity, the conductivity is equal to zero.} \tag{4.13}$$

In the above equations, K_{pl} represents the electrical conductivity of the plastic zone, whereas V_{m}, V_{p}, and V_{pl} represent the volume fraction of the metallic matrix, the porosity, and the plastic zone, respectively.

4.4.2.3. ROMs Model.
Using the ROMs model, the electrical conductivity of the composite materials can also be predicted. This is a simplified approach that takes into consideration the microstructural effects produced by the reinforcement (ceramic

TABLE 4.3. Results of electrical conductivity of MMCs using the various models [36].

MMC	Volume Fraction of Reinforcement (V_r)	Electrical Conductivity ($\times 10^5$) (Ω cm)$^{-1}$			
		Experimental	Rayleigh–Maxwell	ROM	Kerner's
99.5 Al/SiC	0.03	3.01	3.658	3.598	3.524
99.5 Al/C	0.03	3.22	3.648	3.659	3.569

type) on the matrix and the ensuing increase in volume fraction of the electron scattering centers:

$$K_{comp} = K_r V_r + K_{pl} V_{pl} + K_p V_p + K_m V_m \qquad (4.14)$$

In a study by Gupta et al. [36], the results of the electrical conductivity computations using the various models were presented for 99.5 Al/C and 99.5 Al/SiC composites (see Table 4.3). The results revealed that the electrical conductivity of discontinuously reinforced MMCs with addition of low volume fraction of reinforcement cannot be predicted accurately using the Rayleigh–Maxwell equation. This is because the reinforcements change the microstructure of the base matrix. Kerner's model or the simplified ROMs, on the other hand, can be used to predict the electrical conductivity of the MMCs more accurately.

4.4.3. Coefficient of Thermal Expansion

4.4.3.1. ROMs (Upper Bound).
The ROMs model can be used to predict the upper bound of the coefficient of thermal expansion of the composite, α_{comp} [37]:

$$\alpha_{comp} = \alpha_m V_m + \alpha_r V_r \qquad (4.15)$$

where α_m and α_r represent the coefficient of thermal expansion of the matrix material and the reinforcement, respectively.

4.4.3.2. Turner's Model (Lower Bound).
The model by Turner can predict the lower bound of the coefficient of thermal expansion of the composite, α_{comp} [37]. The expression can be written as follows:

$$\alpha_{comp} = \frac{\alpha_m V_m K_m + \alpha_r V_r K_r}{V_m K_m + V_r K_r} \qquad (4.16)$$

where K_m and K_r represent the bulk modulus of the matrix material and the reinforcement, respectively.

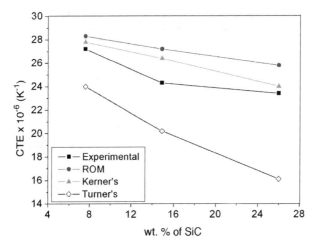

Figure 4.3. Graphical representation showing the experimental and theoretical CTE values derived from the various models, as a function of weight percentage of SiC reinforcement particulates (25 μm) in Mg matrix (data extracted from [38]).

4.4.3.3. Kerner's Model. Another model to predict the coefficient of thermal expansion is proposed by Kerner [35]. The expression is stated below:

$$\alpha_{comp} = \alpha_m - V_r(\alpha_m - \alpha_r)\left[\frac{K_m(3K_r + 4G_m)^2 + (K_r - K_m)(16G_m^2 + 12G_mK_r)}{(4G_m + 3K_r)[4V_rG_m(K_r - K_m) + 3K_rK_m + 4G_mK_m]}\right]$$

(4.17)

where G_m represents the shear modulus of the matrix (Figure 4.3).

4.4.4. Elastic Modulus

4.4.4.1. Rule of Mixtures. The rule of mixtures equation stated below enables the computation of the linear upper bound of composite's elastic modulus, E_{comp} [37]:

$$E_{comp} = \frac{E_m V_m + E_r V_r}{V_m + V_r}$$

(4.18)

where E_{comp}, E_m, and E_r represent the elastic modulus of the composite, the metallic matrix, and the reinforcement, respectively, whereas V_m and V_r represent the volume fraction of the matrix and the reinforcement, respectively.

The nonlinear lower bound of the composite's elastic modulus can be computed by [37]

$$E_{comp} = E_m\left[\frac{E_m V_m + E_r(V_r + 1)}{E_r V_m + E_m(V_r + 1)}\right]$$

(4.19)

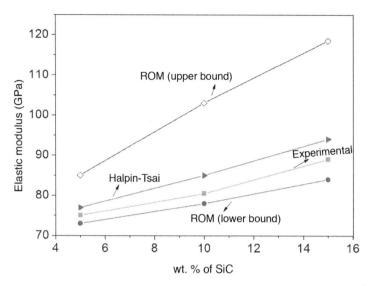

Figure 4.4. Graphical representation showing the experimental and theoretical elastic modulus values derived from the various models, as a function of amount of SiC in aluminum matrix (courtesy: M. Gupta, NUS, Singapore).

4.4.4.2. Halpin–Tsai Model.

The Halpin–Tsai equation [37] is used to compute the elastic modulus of composites with quite a reasonable accuracy:

$$E_{comp} = \frac{E_m(1 + 2sq\,V_r)}{1 - q\,V_r} \tag{4.20}$$

where q can be represented by

$$q = \frac{E_r/E_m - 1}{E_r/E_m + 2s} \tag{4.21}$$

where s is the aspect ratio of the reinforcement.

As shown in Figure 4.4, the experimental results of the elastic modulus were found to be lower than some theoretical values. This can be attributed to the presence of a finite amount of porosity in the metal matrix composite. These porosities adversely affect the elastic response of the composite material because of the reduction in effective load bearing area [39].

4.4.4.3. Effect of Porosity on Elastic Modulus.

A model [40] correlating the elastic modulus of the porous material (E) with the elastic modulus of the fully dense material (E_0) and the volume fraction of the pores (p) is expressed as follows:

$$E = E_0(1 - p^{2/3})^{1.21s} \tag{4.22}$$

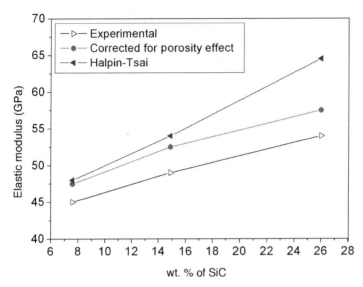

Figure 4.5. Graphical representation showing the experimental and theoretical elastic modulus values derived from the various models, as a function of weight percentage of SiC reinforcement particulates (25 μm) in Mg matrix (data extracted from ref. [38]).

where the power factor s is

$$s = (z/x)^{1/3}\{1 + [(z/x)^{-2} - 1]\cos^2 \alpha_d\}^{1/2} \tag{4.23}$$

where z/x is the aspect ratio of the pores and α_d is the relative orientation of the pores with respect to the stress axis. The orientation factor $\cos^2\alpha_d$ is equivalent to 0.31 for a material with a random orientation of pores which corresponds to $\alpha_d = 56°$ [39] (Figure 4.5).

The results clearly reveal that pore free composites exhibit values that are close to the theoretical predictions and that porosity has a negative effect on the elastic modulus of the materials.

4.4.5. Yield Strength

4.4.5.1. Shear Lag Theories. Attempts have been made to predict the 0.2% yield strength of the metal-based composites. Shear lag theories [41] are perhaps the most widely used but with limited success.

For platelets type of reinforcement, the yield strength of composite, $\sigma_{comp.y}$, is predicted using [41]

$$\sigma_{comp.y} = V_r\sigma_m \left(\frac{s}{4}\right) + V_m\sigma_m \tag{4.24}$$

where V_r and V_m represent the volume fraction of the reinforcement and the matrix material, respectively, whereas σ_m and s represent the yield strength of the matrix material and the aspect ratio of the reinforcement, respectively.

For whiskers/particles type of reinforcement, the yield strength of composite, $\sigma_{comp,y}$, is predicted using [41]

$$\sigma_{comp,y} = \sigma_m \left[\frac{1}{2} V_r(s+2) + V_m \right] \qquad (4.25)$$

4.4.5.2. Strengthening Factors.
Several strengthening mechanisms can contribute to the increase in yield strength of the MMCs:

(i) Generation of geometrically necessary dislocations to accommodate thermal and elastic modulus mismatch between the matrix and reinforcements.
(ii) Load-bearing effects due to the presence of reinforcements.
(iii) Orowan strengthening.
(iv) Hall–Petch effect due to grain size refinement.

The yield strength of the composite, $\sigma_{comp,y}$, can be defined by [42]

$$\sigma_{comp,y} = \sigma_{mo} + \Delta\sigma \qquad (4.26)$$

where σ_{mo} represents the yield strength of the unreinforced matrix, and $\Delta\sigma$ represents the total increment in yield stress of the composite and is estimated by equation (4.27) [1,43]:

$$\Delta\sigma = \sqrt{(\Delta\sigma_{CTE})^2 + (\Delta\sigma_{EM})^2 + (\Delta\sigma_{load})^2 + (\Delta\sigma_{Orowan})^2 + (\Delta\sigma_{Hall-Petch})^2} \qquad (4.27)$$

4.4.5.2.1. THERMAL AND ELASTIC MODULUS MISMATCH. The presence of reinforcements induces an inhomogeneous deformation pattern and high dislocation density in the composite matrix. Hence, composite materials have higher yield strength. $\Delta\sigma_{CTE}$ and $\Delta\sigma_{EM}$ are the stress increment resulted from coefficient of thermal expansion and elastic modulus mismatch between the reinforcements and the metallic matrix, respectively. These two terms can be determined by the Taylor dislocation strengthening relation [42–44]:

$$\Delta\sigma_{CTE} = \sqrt{3}\beta G_m b \sqrt{\rho_{CTE}} \qquad (4.28)$$

and

$$\Delta\sigma_{EM} = \sqrt{3}\alpha G_m b \sqrt{\rho_{EM}}, \qquad (4.29)$$

where α and β are strengthening coefficients, G_m is shear modulus of the matrix, and b is the Burgers vector. ρ_{CTE} and ρ_{EM} represent the dislocation density because of coefficient of thermal expansion and elastic modulus mismatch, respectively.

The increase in dislocation density in the composite matrix can, thus, be attributed to

(i) elastic modulus mismatch and
(ii) coefficient of thermal expansion (CTE) mismatch between the reinforcement and the matrix [45].

The presence of a softer matrix can accommodate the nondeforming, incoherent, and stiffer reinforcements by sufficient generation of geometrically necessary dislocations [46]. The geometrically necessary dislocation density due to elastic modulus mismatch, ρ_{EM}, can be given as [47]

$$\rho_{EM} = \frac{\gamma^m}{b\lambda} \tag{4.30}$$

where γ^m is the shear strain and λ is the local length scale of the deformation field. λ is interpreted to be a distance within which dislocations generated at reinforcements are constrained to move and it is related to inter-reinforcement spacing. λ can be defined as follows [48]:

$$\lambda \approx d_r \left[\left(\frac{1}{2V_r} \right)^{1/3} - 1 \right], \tag{4.31}$$

where d_r is the smallest dimension of the reinforcement and V_r is the volume fraction of the reinforcement.

The geometrically necessary dislocation density due to CTE mismatch, ρ_{CTE}, can be given as [49]

$$\rho_{CTE} = \frac{B V_r \varepsilon_{CTE}}{b(1 - V_r)d_r} \tag{4.32}$$

where B is a geometric constant and ε_{CTE} is the misfit strain due to the different CTE values of the metallic matrix and reinforcements. ε_{CTE} can be expressed as

$$\varepsilon_{CTE} = (C_m - C_r)\Delta T = \Delta C \cdot \Delta T \tag{4.33}$$

where C_m and C_r represent the CTE of the metallic matrix and reinforcements, respectively. ΔT is the temperature change.

4.4.5.2.2. LOAD-BEARING EFFECT. Effective load transfer is highly dependent upon the interfacial bonding between the matrix and the reinforcement. Using the modified shear lag model, the improvement in YS due to load-bearing effect ($\Delta\sigma_{\text{load}}$) can be expressed by [41]

$$\Delta\sigma_{\text{load}} = \frac{V_r\sigma_{\text{mo}}}{2}S_r \tag{4.34}$$

where S_r is the aspect ratio of reinforcement. For the case of particulates as reinforcement, $S_r = 1$ [50]. However, for the case of carbon nanotubes as reinforcement [51],

$$S_r = S\cos^2\theta + \left(\frac{3\pi - 4}{3\pi}\right)\left(1 + \frac{1}{S}\right)\sin^2\theta \tag{4.35}$$

where S is the average aspect ratio of CNT and θ is the misorientation angle of CNT.

4.4.5.2.3. OROWAN STRENGTHENING. Orowan strengthening plays a relatively large role in improving the yield strength of the nanocomposites due to the size of the nanoparticles. With the presence of nanosize reinforcements in the matrix, dislocation loops are formed as dislocation bowed and by-pass the reinforcements. The increment in yield strength due to Orowan strengthening can be described by the Orowan-Ashby equation [48]. The equation can be written as follows:

$$\Delta\sigma_{\text{Orowan}} = \left(\frac{0.13G_mb}{\lambda}\right)\ln\frac{d_r}{2b} \tag{4.36}$$

4.4.5.2.4. HALL–PETCH EFFECT. The contribution to the increase in yield strength due to grain size strengthening can be described by the Hall–Petch equation [52]:

$$\Delta\sigma_{\text{Hall–Petch}} = KD^{-1/2} \tag{4.37}$$

where K is a constant and D is grain size of metallic matrix.

4.4.6. Ductility

Ductility (engineering strain at fracture) of a material is indicative of its ability to plastically deform under load application. Ductility is one of the most intriguing properties which researchers worldwide are attempting to control. Its knowledge can also help the manufacturing engineer to determine the extent of metalworking operations and the design engineer to arrive at more damage tolerant components especially for critical applications.

For a given material, a low ductility value indicates the presence of stress nucleation sites that are capable of generating critical-size flaws for rapid fracture. While a high ductility value reflects that the material's microstructural characteristics are able to tolerate the loading and the extent of localized damages are contained for most part of the loading under the plastic regime.

In the case of MMCs, the eventual failure or plastic deformation capability of the composite materials is governed by the crack initiation and crack growth. The two key sources of crack initiation are

(i) the plastic incompatibility of the reinforcement (e.g., ceramic reinforcement) with the metallic matrix and
(ii) the presence of pores.

Although reinforcement is the essential part of the MMC, the presence of porosity cannot be eliminated regardless of the processing method being adopted [6, 7, 33, 53–57]. Investigators [58, 59] working on MMCs have reported that under tensile loading, cracks will initiate because of high interfacial stresses in the case of metal matrices reinforced with harder and stiffer reinforcement. Once the cracks are initiated, they have the tendency to propagate along the matrix–reinforcement interface according to experimental observations [60].

In general, the addition of ceramic reinforcement deteriorates the ductility of the resultant composite material. However in recent years, with the advancement of nanotechnology and the availability of nanosize reinforcements, contrary results were reported. Magnesium-based composites with the incorporation of nanosize reinforcements exhibited improved ductility [61–63]. Chapter 5 presents a more detailed discussion on the ductility effects of reinforcements for magnesium-based composites. Many studies have been conducted to correlate the role of volume fraction, size, and distribution of reinforcement with ductility.

Cleanliness of the starting materials also plays a crucial role in improving the ductility of MMCs by providing a cleaner matrix–reinforcement interface. Thus, prior to the incorporation of reinforcements into the molten metal, the following can be carried out:

(i) Degassing (to remove adsorbed gases and moisture).
(ii) Surface treatment (to enhance wetting properties) of the ceramic reinforcement.

Mechanical working of the as-processed billets affects the ductility of the composites. Higher reduction results in

(i) lower levels of porosity,
(ii) breaking of the inclusions, and
(iii) more homogeneous distribution of the reinforcements in the MMCs.

All the above factors assist to delay the formation of pores, and thus, indirectly increase the ductility of the MMCs.

In the case of addition of metallic reinforcement, the formation of intermetallic phases, which are more brittle in nature as compared to the metallic matrix, can lead to the reduction in ductility of the resultant composite. Wong et al. [64] reported that with increasing addition of harder Cu reinforcement and increasing formation of brittle intermetallic phase (Mg_2Cu) in the Mg matrix, a decreasing trend of the composite's ductility was observed. Reverse can be expected for metallic reinforcements that do not react with matrix.

4.5. SUMMARY

This chapter gives an overview of the different types of MMCs and the individual roles of the matrix and reinforcement materials, which constitute the metal matrix composite. The interface between the matrix and the reinforcement plays an important part in affecting the overall composite's performance. The performance of the resultant composite is governed by its physical, thermal, electrical, and mechanical properties. Models representing the theoretical prediction of the composite's properties are also presented in this chapter. In essence, the information provided in this chapter reflects that a large number of composites can be designed depending on the nature of end applications, provided existing scientific and technical knowledge corresponding to MMCs is utilized judiciously.

REFERENCES

1. C. S. Goh, J. Wei, L. C. Lee, and M. Gupta (2007) Properties and deformation behavior of Mg-Y_2O_3 nanocomposites. *Acta Materialia*, **55**(15), 5115–5121.
2. N. Chawla and K. K. Chawla (2006). *Metal Matrix Composites*. New York: Springer.
3. P. Babaghorbani, S. M. L. Nai, and M. Gupta. (2009) Development of lead-free Sn-3.5Ag/SnO_2 nanocomposite solders. *Journal of Materials Science: Materials in Electronics*, **20**(6), 571–576.
4. S. M. L. Nai, J. Wei, and M. Gupta (2006) Development of lead-free solder composites containing nanosized hybrid (ZrO_2 + 8 mol.% Y_2O_3) particulates. *Solid State Phenomena*, **111**, 59–62.
5. S. M. L. Nai, J. Wei, and M. Gupta (2006) Improving the performance of lead-free solder reinforced with multi-walled carbon nanotubes. *Materials Science and Engineering A*, **423**(1–2), 166–169.
6. D. J. Lloyd (1994) Particle reinforced aluminum and magnesium matrix composites. *International Materials Reviews*, **39**(1), 1–23.
7. I. A. Ibrahim, F. A. Mohamed, and E. J. Lavernia. Particulate reinforced metal matrix composites—a review. *Journal of Materials Science*, **26**(5), 1137–1156.
8. Website http://www.pa.msu.edu/cmp/csc/ntproperties/ (last accessed May 10, 2008).
9. W. F. Gale and T. C. Totemeier (eds) (2004) *Smithells Metal Reference Book*. Oxford: Elsevier Butterworth-Heinemann.

10. C. J. Smithells (1992) In E. A. Brandes and G. B. Brook (eds), *Metals Reference Book*. London: Butterworth-Heinemann.

11. (1990) *ASM Handbook: Properties and Selection: Non-Ferrous Alloys and Special-Purpose Materials*, Vol. **2**. Materials Park, OH: ASM International, 621 pp.

12. P. Babaghorbani, S. M. L. Nai, and M. Gupta (2008) Development of lead-free nanocomposite solders using oxide based reinforcement. *Proceedings of ASME IMECE 2008, 31 Oct–6 Nov 2008*, Boston, USA.

13. R. Morrell (1985) *Handbook of Properties of Technical & Engineering Ceramics*. London: HMSO.

14. Website: http://www.ceramics.nist.gov/srd/summary/Y2O3.htm (last accessed May 10, 2008).

15. J. F. Douglas, J. M. Gasiorek, and J. A. Swaffield (1994) *Fluid Mechanics*, 3rd edition. New York: John Wiley & Sons, Ltd., 367 pp.

16. S. M. L. Nai and M. Gupta (2002) Synthesis of Al/SiC based functionally gradient materials using technique of gradient slurry disintegration and deposition: effect of stirring speed. *Materials Science and Technology*, **18**(6), 633–641.

17. M. Gupta and C. Y. Loke (2000) Synthesis of free standing, one dimensional, Al-SiC based functionally gradient materials using gradient slurry disintegration and deposition. *Materials Science and Engineering A*, **276**, 210–217.

18. G. N. Hassold, E. A. Holm, and D. J. Srolovitz (1990) Effects of particle size on inhibited grain growth. *Scripta Metallurgica et Materialia*, **24**, 101–106.

19. R. E. Johnson, Jr (1959) Conflicts between Gibbsian thermodynamics and recent treatments of interfacial energies in solid-liquid-vapor. *The Journal of Physical Chemistry*, **63**(10), 1655–1658.

20. F. Delannay, L. Froyen, and A. Deruyttere (1987) The wetting of solids by molten metals and its relation to the preparation of metal-matrix composites. *Journal of Materials Science*, **22**(1), 1–16.

21. A. Mortensen, J. A. Cornie, and M. C. Flemings (1988) Solidification processing of metal-matrix composites. *Journal of Metals*, **40**, 12–19.

22. R. L. Mehan and E. Feingold (1967) Room and elevated temperature strength of α-Al$_2$O$_3$ whiskers and their structural characteristics. *Journal of Materials*, **2**, 239–270.

23. K. K. Chawla (1987) *Composite Materials: Science and Engineering*. New York: Springer-Verlag.

24. J. M. Howe (1993) Bonding structure, and properties of metal/ceramic interfaces. I: Chemical bonding, chemical reaction, and interfacial structure. *International Materials Review*, **38**(5), 233–256.

25. M. Ruhle (1993) Structure and chemistry of metal/ceramic interfaces. In S. Suresh, A. Mortensen, and A. Needleman (eds) *Fundamentals of Metal-Matrix Composites*. Boston, MA: Butterworth-Heinemann, pp. 81–108.

26. S. Ochiai and Y. Murakami (1979) Tensile strength of composites with brittle reaction zones at interfaces. *Journal of Materials Science*, **14**(4), 831–840.

27. L. M. Tham, L. Cheng, and M. Gupta (2001) Effect of limited matrix–reinforcement interfacial reaction on enhancing the mechanical properties of aluminum–silicon carbide composites. *Acta Materialia*, **49**(16), 3243–3253.

28. N. Eustathopoulos and A. Mortensen (1993) Capillary phenomena, interfacial bonding, and reactivity. In S. Suresh, A. Mortensen, and A. Needleman (eds) *Fundamentals of Metal-Matrix Composites*. Boston, MA: Butterworth-Heinemann, pp. 42–58.

29. A. Mortensen and I. Jin (1992) Solidification processing of metal matrix composites. *International Materials Reviews*, **37**(3), 101–128.

30. A. Needleman, S. R. Nutt, S. Suresh, and V. Tvergaard (1993) Matrix, reinforcement, and interfacial failure. In S. Suresh, A. Mortensen, and A. Needleman (eds) *Fundamentals of Metal-Matrix Composites*. Boston, MA: Butterworth-Heinemann, pp. 233–250.

31. S. R. Nutt and J. M. Duva (1986) A failure mechanism in Al-SiC composites. *Scripta Metallurgica*, **20**(7), 1055–1058.

32. S. R. Nutt and A. Needleman (1987) Void nucleation at fiber ends in Al-SiC composites. *Scripta Metallurgica*, **21**(5), 705–710.

33. A. L. Geiger and J. A. Walker (1991) The processing and properties of discontinuously reinforced aluminum composites. *JOM*, **43**(8), 8–15.

34. Lord Rayleigh (1892) On the influence of obstacles arranged in rectangular order upon the properties of a medium. *Philosophical Magazine of Journal Science*, **34**, 481–502.

35. E. H. Kerner (1956) The Electrical Conductivity of Composite Media. *Proceedings of the Physical Society of London*, **B69**, 802–807.

36. M. Gupta, G. Karunasiri, and M. O. Lai (1996) Effect of presence and type of particulate reinforcement on the electrical conductivity of non-heat treatable aluminum. *Materials Science and Engineering*, **A219**, 133–141.

37. W. L. E. Wong, S. Karthik, and M. Gupta (2005) Development of high performance Mg-Al$_2$O$_3$ composites containing Al$_2$O$_3$ in submicron length scale using microwave assisted rapid sintering. *Materials Science and Technology*, **21**(9), 1063–1070.

38. S. C. V. Lim, M. Gupta, and L. Lu (2001) Processing, microstructure, and properties of Mg–SiC composites synthesised using fluxless casting process. *Materials Science and Technology*, **17**(7), 823–832.

39. G. E. Fougere, L. Riester, M. Ferber, J. R. Weertman, and R. W. Siegel (1995) Young's modulus of nanocrystalline Fe measured by nanoindentation. *Materials Science and Engineering A*, **204**, 1–6.

40. A. R. Boccaccini, G. Ondracek, P. Mazilu, and D. Windelberg (1993) On the effective Young's modulus of elasticity for porous materials: Microstructure modeling and comparison between calculated and experimental values. *Journal of the Mechanical Behavior of Materials*, **4**(2), 119–128.

41. V. C. Nardone and K. M. Prewo (1986) On the strength of discontinuous silicon carbide reinforced aluminum composites. *Scripta Metallurgica*, **20**(1), 43–48.

42. L. H. Dai, Z. Ling, and Y. L. Bai (2001) Size-dependent inelastic behavior of particle-reinforced metal-matrix composites. *Composites Science and Technology*, **61**(8), 1057–1063.

43. T. W. Clyne and P. J. Withers (1993) *An Introduction to Metal Matrix Composites*. Cambridge: Cambridge University Press.

44. C. S. Goh, J. Wei, L. C. Lee, and M. Gupta (2006) Development of novel carbon nanotube reinforced magnesium nanocomposites using the powder metallurgy technique. *Nanotechnology*, **17**, 7–12.

45. M. F. Ashby (1970) The deformation of plastically non-homogenous crystals. *Philosophical Magazine*, **21**, 399–424.

46. T. R. McNelley, G. R. Edwards, and O. D. Sherby (1977) A microstructural correlation between the mechanical behavior of large volume fraction particulate composites at low and high temperatures. *Acta Metallurgica*, **25**, 117–124.

47. M. Kouzeli and A. Mortensen (2002) Size dependent strengthening in particle reinforced aluminum. *Acta Materialia*, **50**(1), 39–51.

48. Z. Zhang and D. Chen (2006) Consideration of Orowan strengthening effect in particulate-reinforced metal matrix nanocomposites: A model for predicting their yield strength. *Scripta Materialia*, **54**(7), 1321–1326.

49. R. J. Arsenault and N. Shi (1986) Dislocations generation due to differences between the coefficients of thermal expansion. *Materials Science and Engineering*, **81**(1–2), 175–187.

50. N. Ramakrishnan (1996) An analytical study on strengthening of particulate reinforced metal matrix composites. *Acta Materialia*, **44**(1), 69–77.

51. H. J. Ryu, S. I. Cha, and S. H. Hong (2003) Generalized shear-lag model for load transfer in SiC/Al metal-matrix composites. *Journal of Materials Research*, **18**(12), 2851–2858.

52. X. L. Zhong, W. L. E. Wong, and M. Gupta (2007) Enhancing strength and ductility of magnesium by integrating it with aluminum nanoparticles. *Acta Materialia*, **55**(18), 6338–6344.

53. M. Gupta, L. Lu, M. O. Lai, and S. E. Ang (1995) Effects of type of processing on the microstructural features and mechanical properties of Al-Cu/SiC metal matrix composites. *Materials and Design*, **16**(2), 75–81.

54. M. Gupta, M. O. Lai, and C. Y. Soo (1996) Effect of type of processing on the microstructural features and mechanical properties of Al-Cu/SiC metal matrix composites. *Materials Science and Engineering A*, **210**, 114–122.

55. D. L. McDanels (1985) Analysis of stress-strain, fracture, and ductility behavior of aluminum matrix composites containing discontinuous silicon carbide reinforcement. *Metallurgical and Materials Transactions A*, **16**(6), 1105–1115.

56. W. L. E. Wong, S. Karthik, and M. Gupta (2005) Development of hybrid Mg/Al$_2$O$_3$ composites with improved properties using microwave assisted rapid sintering route. *Journal of Materials Science*, **40**(13), 3395–3402.

57. M. Gupta, M. O. Lai, and D. Saravanaranganathan (2000) Synthesis, microstructure and properties characterization of disintegrated melt deposited Mg/SiC composites. *Journal of Materials Science*, **35**(9), 2155–2165.

58. S. Qin and M. Gupta (1995) The minimum volume fraction of SiC reinforcement required for strength improvement of an Al-Cu based composite. *Journal of Materials Science*, **30**(20), 5223–5227.

59. Z. Xiao (1991) In: Ph.D Dissertation. New Jersey: Rutgers University.

60. Y. Wu and E. J. Lavernia (1992) Strengthening behavior of particulate reinforced MMCs. *Scripta Metallurgica et Materialia*, **27**(2), 173–178.

61. C. S. Goh, J. Wei, L. C. Lee, and M. Gupta (2006) Simultaneous enhancement in strength and ductility by reinforcing magnesium with carbon nanotubes. *Materials Science and Engineering A*, **423**, 153–156.

62. S. F. Hassan and M. Gupta (2007) Development and characterization of ductile Mg/Y_2O_3 nanocomposites. *Transactions of the ASME*, **129**, 462–467.

63. Q. B. Nguyen and M. Gupta (2008) Increasing significantly the failure strain and work of fracture of solidification processed AZ31B using nano-Al_2O_3 particulates. *Journal of Alloys and Compounds*, **459**, 244–250.

64. W. L. E. Wong and M. Gupta (2007) Development of Mg/Cu nanocomposites using microwave assisted rapid sintering. *Composites Science and Technology*, **67**(7–8), 1541–1552.

5

MAGNESIUM COMPOSITES

This chapter focuses on magnesium matrix composites. It presents the overview of the factors influencing the end properties of resultant magnesium-based composites. In particular, the effects of matrix (pure magnesium or magnesium alloy) and reinforcements (type, shape, amount, and length scale of reinforcements) on the properties of the magnesium composites are presented. The ductility effects of the reinforcements are also discussed here. Lastly, this chapter is dedicated to introduce the various magnesium-based composites containing different reinforcements and their end properties. This chapter can serve as a good magnesium composite material selection guide for engineers, scientists, technicians, teachers, and students in the fields of materials design, development and selection, manufacturing, and engineering.

5.1. INTRODUCTION

Magnesium composites are reportedly synthesized using several approaches as earlier presented in Chapter 2. The end properties of the magnesium composites are dependent on several factors, namely, the processing method, type of matrix alloy, type of reinforcement, shape of reinforcement, amount of reinforcement, length scale of reinforcement,

Magnesium, Magnesium Alloys, & Magnesium Composites, by Manoj Gupta and Nai Mui Ling, Sharon
© 2010 John Wiley & Sons, Inc.

and heat treatment. Different combination of these factors will result in composites of different end properties, which can serve different applications.

The following sections briefly describe how these influencing factors attribute to the end properties of the resultant magnesium composites.

5.2. MATERIALS

5.2.1. Matrix

Magnesium is one of the metallic materials with the lowest density and is, thus, widely used as a matrix material because of its high specific strength. In existing literature, both pure magnesium and commercial grade magnesium alloys are chosen as the matrix material for composites. In the case of commercial grade magnesium alloys, AZ91 and AZ31 are two of the most commonly used matrix materials for the synthesis of magnesium-based composites.

5.2.2. Reinforcements

For the synthesis of composite materials, reinforcements are intentionally introduced into the magnesium matrix. The reinforcements of different types, shapes, amounts, and length scales (micrometer size, sub-micrometer size, and nanosize), are judiciously selected to enhance the end properties of the base magnesium material, so as to serve a specific application.

5.2.2.1. Type of Reinforcements.   For magnesium matrix composites, the reinforcements intentionally incorporated into the magnesium matrix can be grouped into the following (Table 5.1):

(a) Ceramic
(b) Metallic
(c) Intermetallic

The ceramic type of reinforcements is the most commonly used in magnesium-based composites. They range from carbides, borides, or oxides. Of all the ceramic reinforcements, the most widely investigated and used reinforcing ceramics with pure magnesium and commercial grade magnesium alloys is the SiC particulates. It is thermodynamically stable in many molten Mg alloys and it has relatively good wettability with Mg in comparison to the other ceramic reinforcements [1, 2].

For the case of metallic reinforcements, elemental metal powders such as Cu, Ni, Ti, Mo, and Al have been reportedly used [72–84].

Since their discovery, carbon nanotubes (CNTs) have attracted much attention as a potential reinforcement candidate for high strength, high performance, and light-weight composites, owing to their superior strength (up to 150 GPa) and Young's modulus (up to 1 TPa) [69, 93].

TABLE 5.1. List of some reinforcements used in magnesium matrix composites [1–92].

Type of Reinforcement	Name of Reinforcement	Symbol
Ceramic	Boron carbide	B_4C
	Silicon carbide	SiC
	Titanium carbide	TiC
	Alumina, Aluminum oxide	Al_2O_3
	Magnesium oxide	MgO
	Tin oxide	SnO_2
	Yttria, Yttrium oxide	Y_2O_3
	Zirconia, Zirconium oxide	ZrO_2
	Titanium boride	TiB_2
	Zirconium boride	ZrB_2
Metallic	Aluminum	Al
	Copper	Cu
	Molybdenum	Mo
	Nickel	Ni
	Titanium	Ti
Others	Carbon	C
	Carbon nanotubes	CNTs

In a study by Hassan et al. [94], different types of nanosize oxide particulates (namely, Al_2O_3, Y_2O_3, and ZrO_2) were, respectively, introduced into pure magnesium. The grain size measurement results (refer to Table 5.2) revealed that with the addition of nanosize particulates into the magnesium matrix, grain refinement was observed. This was attributed to the following factors:

(i) Ability of the presence of fine second-phase particulates to nucleate the magnesium grains during recrystallization.

(ii) Inhibition of growth of recrystallized magnesium grains due to pinning by the finer reinforcement particulates.

TABLE 5.2. Results of mechanical properties of Mg and Mg-based composites (reinforced with different types of nanosize oxide particulates), synthesized using the blend-press-sinter type powder metallurgy technique [94].

Material	Mg	Mg/1.1 vol.% Al_2O_3	Mg/1.1 vol.% Y_2O_3	Mg/1.1 vol.% ZrO_2
Size of oxide particulates	—	50 nm (avg.)	29 nm (avg.)	29–68 nm
Grain size (μm)	60 ± 10	31 ± 13	12 ± 3	11 ± 3
Aspect ratio of grain	1.6 ± 0.3	2.2 ± 1.3	1.6 ± 0.4	1.5 ± 0.2
Microhardness (HV)	37.4 ± 0.4	69.5 ± 0.5	51.0 ± 0.7	45.7 ± 0.6
Marcohardness (15 HRT)	43.5 ± 0.3	59.7 ± 0.5	49.0 ± 0.5	48.4 ± 0.7
0.2% Yield strength (MPa)	132 ± 7	194 ± 5	153 ± 3	146 ± 1
Tensile strength (MPa)	193 ± 2	250 ± 3	195 ± 2	199 ± 5
Ductility (%)	4.2 ± 0.1	6.9 ± 1.0	9.1 ± 0.2	10.8 ± 1.3

The results also showed that the addition of Y_2O_3 and ZrO_2 particulates led to more prominent grain refinement than for the case of Al_2O_3 particulates. The likely reason is their higher thermal stability in magnesium [94–97].

The mechanical properties (in terms of microhardness, macrohardness, 0.2% yield strength, and tensile strength) of the various composites all yielded improvement over their monolithic counterpart (see Table 5.2). The most significant enhancement in mechanical properties is observed for the case of Mg/Al_2O_3 composite. This could be attributed to the matrix/reinforcement interfacial compatibility, as Al_2O_3 is know to be the most vulnerable to diffusion-controlled superficial reaction with magnesium [94–96, 98]. Hence, strong interfacial bonding is formed between Mg and Al_2O_3 and this translates to considerable strengthening effect.

5.2.2.2. Shape of Reinforcements.   As presented earlier in Section 4.2, there are various forms of reinforcement:

   (i) Particulates
  (ii) Short fibers or whiskers
 (iii) Continuous fiber
  (iv) Laminates or sheets
   (v) Interconnected reinforcement
  (vi) Singular metal core reinforcement

Studies have shown that the shape of reinforcements plays a critical role to influence the mechanical properties of the composite. Rod-shape reinforcements like carbon nanotubes are reported to be more effective in impeding dislocation motion and strengthening the metallic matrix than spherical-shaped reinforcements. On the other hand, angular-shaped particulates can act as local stress raisers, reducing the ductility of the resultant magnesium composite.

In a study by Kelly *et al.* [99], the rod-shaped reinforcements resulted in almost twice as much strengthening than the spherical-shaped reinforcements of the same volume fraction.

5.2.2.3. Amount of Reinforcements.   With the addition of increasing amount of reinforcements, there is either an increasing trend or decreasing trend in the composite's properties. In the case of the mechanical properties, typically, with the increasing amount of reinforcements, the composite's hardness (in terms of micro- and macrohardness) and strength (in terms of 0.2% yield strength and ultimate tensile strength) values increase correspondingly. However, often there is a threshold amount whereby further addition of reinforcements will be detrimental to the material's mechanical properties.

The influence of the amount of reinforcements on the properties of a variety of magnesium-based composites is described in Sections 5.3–5.10.

5.2.2.4. Length Scale of Reinforcements.   The advancement of reinforcement fabrication technology has led to the progressive availability of reinforcements in

TABLE 5.3. Results of mechanical properties of Mg and Mg/Al$_2$O$_3$ composites (reinforced with different length scale of Al$_2$O$_3$), synthesized using the blend-press-sinter type powder metallurgy technique [10].

Material	Mg	Mg/Al$_2$O$_3$	Mg/Al$_2$O$_3$	Mg/Al$_2$O$_3$
Amount of Al$_2$O$_3$ (vol.%)	—	1.1	1.1	1.1
Size of Al$_2$O$_3$	—	50 nm	0.3 μm	1.0 μm
Interparticle spacing (μm)	—	0.47	2.85	9.49
Grain size (μm)	60 $\pm$ 10	31 $\pm$ 13	11 $\pm$ 4	11 $\pm$ 3
Porosity (%)	0.08	0.09	0.01	0.01
Microhardness (HV)	37.4 $\pm$ 0.4	69.5 $\pm$ 0.5	51.8 $\pm$ 0.3	51.2 $\pm$ 0.5
Marcohardness (15 HRT)	43.5 $\pm$ 0.3	59.7 $\pm$ 0.5	56.3 $\pm$ 0.5	50.3 $\pm$ 0.5
0.2% Yield strength (MPa)	132 $\pm$ 7	194 $\pm$ 5	182 $\pm$ 3	172 $\pm$ 1
Tensile strength (MPa)	193 $\pm$ 2	250 $\pm$ 3	237 $\pm$ 1	227 $\pm$ 2
Ductility (%)	4.2 $\pm$ 0.1	6.9 $\pm$ 1.0	12.1 $\pm$ 1.4	16.8 $\pm$ 0.4
Work of fracture (J/m^3)	7.1 $\pm$ 0.3	15.5 $\pm$ 2.6	25.0 $\pm$ 3.3	34.7 $\pm$ 0.8
$\sigma_{0.2\% \text{ YS}}/\rho^a$	76	110	103	97
$\sigma_{\text{UTS}}/\rho^a$	111	142	134	129

[a]Theoretical density (ρ) = 1.8782 g cm^{-3} (assuming zero porosity).

micrometer size, sub-micrometer size, and even nanosize. This motivated the research community and industry to adopt these different length scale reinforcements in the fabrication of metal matrix composites.

Hassan and his coworkers [10] investigated the influence of length scale of Al$_2$O$_3$ particulates on the properties of elemental magnesium (see Table 5.3). The composite materials were synthesized using the powder metallurgy method. The results of the grain size study showed that the addition of micrometer-size and sub-micrometer-size Al$_2$O$_3$ particulates was more effective in grain refinement than the addition of nanosize particulates. This observation could be due to the clustering tendency of nanosize particulates [100, 101]. Furthermore, it was evident from the mechanical test results (in terms of hardness and tensile test results) that the magnesium composites with the nanosize Al$_2$O$_3$ particulates yielded the best mechanical properties over their monolithic Mg and also those reinforced with micrometer-size and sub-micrometer-size particulates. The following are the attributing factors to the improved mechanical performance with the incorporation of nanosize particulates:

(i) Decrease in interparticle spacing due to increase in number of particulate obstacle in dislocation movement.

(ii) Increase in density of thermally induced dislocation with the decrease in size of reinforcement.

Interestingly, it was reported that an increasing particulate size (from 50 nm to 1.0 μm) led to the increase in ductility of pure magnesium (see Table 5.3). However, with the addition of nanosize Al$_2$O$_3$ particulates, their influence on the increase of ductility was not as significant as their micrometer-size and sub-micrometer-size counterparts. This finding was attributed to

(i) the relatively higher porosity level and
(ii) the increase in clustering tendency of the nanosize particulate, which resulted in clustering-related porosity.

The specific strength values of the composite with nanosize particulates are also superior (see Table 5.3). Thus, such composite system is most suitable for strength-based designs that require higher yield strength and tensile strength. On the other hand, composite systems with additions of coarser particulate sizes are deemed more suitable for damage tolerant designs with good formability because of their higher work of fracture values.

5.2.2.5. Ductility Effects of Reinforcements.
Magnesium due to its hexagonal close-packed (HCP) structure possesses only three independent slip systems and, thus, has the disadvantage of limited ductility. In order to improve the ductility of pure magnesium, alloying elements were strategically incorporated into the magnesium matrix to form magnesium alloys. In recent years, studies [7–71] have shown that the addition of nanosize discontinuous reinforcements (such as Al_2O_3, MgO, Y_2O_3, ZrO_2, SiC, and CNTs) in certain amounts into the magnesium-based matrix can also result in a simultaneous increase in the overall composite's strength and ductility. However, it was noted that apart from the type and the size of reinforcement, the composite synthesis method also plays a role in influencing the composite's ductility value.

Table 5.4 lists the improvement in ductility of some magnesium-based composites. In the case of magnesium-based composites reinforced with nanosize Al_2O_3 particulates, composites synthesized with both powder metallurgy method and disintegrated melt deposition (liquid metallurgy) method yielded composites with improved ductility compared with their monolithic counterparts. However, for the case of magnesium-based composites reinforced with nanosize Y_2O_3 and ZrO_2 particulates, only composites fabricated using the powder metallurgy method resulted in enhanced ductility. Those composites fabricated using the disintegrated melt deposition (liquid metallurgy) method exhibited a decreasing trend in ductility (see Table 5.4). For Al_2O_3 particulates, it was reported that using the powder metallurgy method, the magnesium-based composites reinforced with micrometer-size, sub-micrometer-size, and nanosize Al_2O_3 particulates, all exhibited enhanced ductility and strength values (see Tables 5.3 and 5.4).

Magnesium-based composites reinforced with CNTs and fabricated using either the powder metallurgy (blending followed by sintering) or the disintegrated melt deposition (liquid metallurgy) method [65–67, 70] yielded composites with improved ductility values over their monolithic counterparts as shown in Table 5.4. However, it was also noted that for AZ91D-CNT composites synthesized using the mechanical milling method exhibited a significant decrease in ductility values with an increase in 0.2% proof stress and tensile stress [68].

TABLE 5.4. Ductility improvement results of different composite systems.

Reinforcement			Type of Matrix	Composite Synthesis Method	Increase in Ductility or Failure Strain	Reference
Type	Size	Amount				
SiC	50 nm	1.5 wt%	Mg4Zn	Melt casting (ultrasonic cavitation)	135% (D)	[45]
SiC	50 nm	0.5 wt%	Pure Mg	Melt casting (ultrasonic cavitation)	11% (D)	[44]
SiC	50 nm	1.5 wt%	Mg6Zn	Melt casting (ultrasonic cavitation)	3% (D)	[46]
SiC	45–55 nm	1.84 wt%	Pure Mg	Powder metallurgy (microwave sintering)	31% (D)	[47]
SiC	50 nm	2.5 wt% (1.1 vol.%)	Pure Mg	Disintegrated melt deposition	89% (D)	[7,9]
Al$_2$O$_3$	50 nm	1.5 vol.%	AZ31B	Disintegrated melt deposition	427% (FS)	[13]
Al$_2$O$_3$	50 nm	1.5 vol.%	AZ31	Disintegrated melt deposition	115% (FS)	[13]
Al$_2$O$_3$	50 nm	1.1 vol.%	Pure Mg	Powder metallurgy (conventional sintering)	64% (D)	[8, 10]
Al$_2$O$_3$	0.3 μm	1.1 vol.%	Pure Mg	Powder metallurgy (conventional sintering)	188% (D)	[10]
Al$_2$O$_3$	1.0 μm	1.1 vol.%	Pure Mg	Powder metallurgy (conventional sintering)	300% (D)	[10]
Y$_2$O$_3$	29 nm	0.6 wt%	Pure Mg	Powder metallurgy (conventional sintering)	285% (D)	[56]
Y$_2$O$_3$	30–50 nm	2.0 wt% (0.7 vol.%)	Pure Mg	Powder metallurgy (microwave sintering)	15% (D)	[58]
Y$_2$O$_3$	30–50 nm	0.7 vol.%	Pure Mg	Powder metallurgy (microwave sintering)	61% (FS)	[61]
Cu	25 nm	0.3 vol.%				
Y$_2$O$_3$	30–50 nm	0.7 vol.%	Pure Mg	Powder metallurgy (microwave sintering)	38% (FS)	[62]
Ni	20 nm	0.6 vol.%				
ZrO$_2$	29–68 nm	1.9 wt% (0.66 vol.%)	Pure Mg	Powder metallurgy (conventional sintering)	171% (D)	[64]
Ti	19 μm	5.6 wt% (2.2 vol.%)	Pure Mg	Disintegrated melt Deposition	44% (D)	[79]
Mo	40 μm	3.6 wt% (0.63 vol.%)	Pure Mg	Disintegrated melt deposition	53% (D)	[82]
Al	18 nm	0.76 wt% (0.5 vol.%)	Pure Mg	Powder metallurgy (conventional sintering)	35% (FS)	[84]
CNT	20 nm (diameter)	1.3 wt%	Pure Mg	Disintegrated melt deposition	69% (D)	[65, 66]
CNT	20 nm (diameter)	0.18 wt%	Pure Mg	Powder metallurgy (conventional sintering)	22% (D)	[65, 67]
CNT	40–70 nm (diameter)	1.0 vol.%	AZ31	Disintegrated melt deposition	68% (FS)	[70]

D, ductility; FS, failure strain.

Moreover, with the addition of micrometer-size Ti and Mo metallic reinforcements, their respective composite systems fabricated using the disintegrated melt deposition method exhibited enhanced ductility compared with their unreinforced counterparts [79, 82]. In contrast, addition of micrometer-size Cu [72, 73] and Ni [77, 78] metallic reinforcements in Mg failed to improve their respective composites' ductility, despite exhibiting an improvement in strength.

The increment in ductility of the magnesium matrix by the incorporation of reinforcements may be attributed to the activation of a nonbasal slip system [67, 71]. The fractographs of the composites that exhibited improved ductility also revealed a mixed mode type of failure with increased levels of plastic deformation compared with the unreinforced Mg, which showed predominantly brittle mode of fracture.

A study by Goh *et al.* [66, 71] on Mg-CNT composites fabricated using the disintegrated melt deposition (liquid metallurgy) method reported an increase in ductility in Mg reinforced with up to 1.6 wt% of multiwalled CNTs (MWCNTs). With the addition of 1.3 wt% of CNTs, the ductility of the corresponding Mg-CNT nanocomposite improved most significantly by 69% (see Table 5.4). The improvement in ductility was found to be because of the high activity of the basal slip system and the initiation of prismatic $\langle a \rangle$ slip. It was reported that during the extrusion process, the CNTs present in the Mg matrix aided in the activation of the prismatic and cross-slip in the matrix. Furthermore, the results of the texture analysis showed the alignment of the basal planes along the extrusion direction. This elucidates that during the tensile deformation, both the basal and nonbasal slips are activated in monolithic Mg and Mg–CNT composite.

In another study by Paramsothy *et al.* [70], 1.0 vol.% of CNT was incorporated into AZ31 magnesium alloy using the disintegrated melt deposition (liquid metallurgy) method. It was reported that compared with the unreinforced AZ31 alloy, the tensile failure strain improved by 68% in the case of AZ31/1.0 vol.% CNT. This finding can be attributed to AZ31/1.0 vol.% CNT:

(i) exhibiting (1 0-1 1) dominant texture in the transverse and longitudinal directions (or (0 0 0 2) basal plane at about 45° angle to force axis),

(ii) having reasonably uniform distribution of CNT, and

(iii) having presence of smaller and more rounded intermetallic particles (see Table 5.5).

As reported in other studies on composites reinforced with nanoparticles [22, 55], the presence of the dispersed nanoparticles in the brittle Mg matrix serve to

(i) provide sites where cleavage cracks are opened ahead of the advancing crack front,

(ii) dissipate the stress concentration that would otherwise exist at the crack front, and

TABLE 5.5. Results of the characteristics of intermetallic particles present in the AZ31 and AZ31/1.0 vol.% CNT [70].

Material	AZ31	AZ31/1.0 vol.% CNT
Size of intermetallic particle[a] (μm)	3.3 ± 1.1	0.7 ± 0.3
Roundness ratio of intermetallic particle[a, b]	1.9	1.2

[a] Based on approximately 100 particles.
[b] Roundness measures the sharpness of a particle's edges and corners expressed by (perimeter)^2/4π (area) [21].

(iii) change the local effective stress state from plane strain to plane stress in the neighborhood of crack tip.

The presence of CNT assists in breaking down the β-Al$_{12}$Mg$_{17}$ intermetallic phase. The smaller intermetallic particles located at the grain boundaries, coupled with the change in their distribution from a predominantly aggregated type to dispersed type as observed in the study, both could aid to enhance the material's ductility.

Furthermore, it was also reported that the addition of 1.0 vol.% CNT resulted in grain refinement (see Table 5.6). Grain refinement assists hexagonal metals in ductility increment where limited intergranular fracture arises from intercrystalline stresses [102].

5.3. MAGNESIUM-BASED COMPOSITES WITH Al₂O₃

In this section, properties of magnesium-based composites containing Al$_2$O$_3$ as reinforcement are presented. For convenience, details of processing type (powder metallurgy or solidification), nature of starting materials, and type of secondary processing are presented. Results reveal that properties of base matrix (Mg or its alloy) is governed by size of reinforcement, amount of reinforcement, distribution of reinforcement, and whether it is hybridized with the same reinforcement at different length scale or different type of reinforcement such as CNTs.

TABLE 5.6. Results of the characteristics of grain present in the AZ31 and AZ31/1.0 vol.% CNT [70].

Material	AZ31	AZ31/1.0 vol.% CNT
Size of grain[a] (μm)	4.0 ± 0.9	2.7 ± 0.7
Aspect ratio of grain[a]	1.4	1.4

[a] Based on approximately 100 grains.

5.3.1. Addition of Sub-Micrometer-Size Al₂O₃

5.3.1.1. Mg Reinforced with 0.3 μm Al₂O₃ (by Disintegrated Melt Deposition)

Processing method	Disintegrated melt deposition
Matrix material used	Pure Mg (>99.9% purity)
Size of Al₂O₃ used	0.3 μm
Extrusion ratio	20.25:1
Extrusion temperature	250°C

Refer to Table 5.7.

TABLE 5.7. Characteristics of Mg and Mg/Al₂O₃ [17].

Material	(1) Mg	(2) Mg/0.7 Al₂O₃	(3) Mg/1.1 Al₂O₃	(4) Mg/2.5 Al₂O₃
Wt% of Al₂O₃ (vol.%)	—	1.5 (0.7)	2.5 (1.1)	5.5 (2.5)
Density (g/cm³)	1.7397	1.7541	1.7636	1.7897
Porosity (%)	0.02	0.04	0.06	0.32
0.2% YS (MPa)	97 ± 2	214 ± 4	200 ± 1	222 ± 2
UTS (MPa)	173 ± 1	261 ± 5	256 ± 1	281 ± 5
Ductility (%)	7.4 ± 0.2	12.5 ± 1.8	8.6 ± 1.1	4.5 ± 0.5
Work of fracture (J/m³)	11.1 ± 0.3	28.9 ± 4.7	20.9 ± 2.8	10.0 ± 1.3

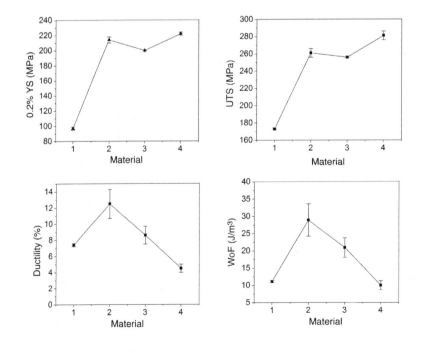

5.3.1.2. Mg Reinforced with 0.3 μm Al₂O₃ (by Powder Metallurgy—Microwave Sintering)

Processing method	Powder metallurgy route
Sintering method	Hybrid microwave-assisted sintering
Matrix material used	Pure Mg (98.5% purity)
Size of Mg used	60–300 μm
Size of Al$_2$O$_3$ used	0.3 μm
Sintering time	25 min
Extrusion temperature	350°C
Extrusion ratio	25:1

Refer to Table 5.8.

TABLE 5.8. Characteristics of Mg and Mg/Al$_2$O$_3$ [18].

Material	Mg	Mg/2.5 Al$_2$O$_3$	Mg/5.0 Al$_2$O$_3$
Wt% of Al$_2$O$_3$ (vol.%)	–	5.5 (2.5)	10.7 (5.0)
Density (g/cm^3)	1.739	1.763	1.820
Porosity (%)	0.07	1.83	1.66
Elastic modulus (GPa)	45.0 ± 0.5	48.8 ± 0.9	58.1 ± 1.6
0.2% YS (MPa)	116.6 ± 11.1	130.3 ± 3.7	158.5 ± 9.7
UTS (MPa)	168.1 ± 10.6	167.3 ± 6.6	214.2 ± 8.8
Ductility (%)	9.0 ± 0.3	3.9 ± 0.1	2.8 ± 0.3

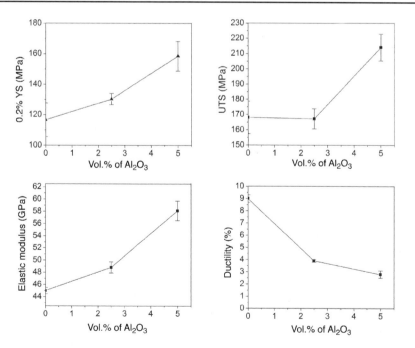

5.3.2. Addition of Nanosize Al_2O_3

5.3.2.1. Mg Reinforced with 50 nm Al_2O_3 (by Disintegrated Melt Deposition)

Processing method   Disintegrated melt deposition
Matrix material used   Pure Mg ($>99.9\%$ purity)
Size of Al_2O_3 used   50 nm
Extrusion ratio   20.25:1
Extrusion temperature   250°C

Refer to Table 5.9.

TABLE 5.9. Characteristics of Mg and Mg/Al_2O_3 [7, 9].

Material	Mg	Mg/0.22 Al_2O_3	Mg/0.66 Al_2O_3	Mg/1.11 Al_2O_3
Wt% of Al_2O_3 (vol.%)	—	0.5 (0.22)	1.5 (0.66)	2.5 (1.11)
Density (g/cm^3)	1.7397	1.7436	1.7476	1.7623
Porosity (%)	0.02	0.07	0.44	0.14
CTE ($\times 10^{-6}$/°C)	28.4 ± 0.3	27.5 ± 0.1	28.0 ± 0.9	25.1 ± 0.3
Dynamic elastic modulus (GPa)	42.8	44.6	45.3	52.7
0.2% YS (MPa)	97 ± 2	146 ± 5	170 ± 4	175 ± 3
UTS (MPa)	173 ± 1	207 ± 11	229 ± 2	246 ± 3
Ductility (%)	7.4 ± 0.2	8.0 ± 2.3	12.4 ± 2.1	14.0 ± 2.4
Work of fracture (J/m^3)	11.1 ± 0.3	23.2 ± 11.7	30.0 ± 3.2	31.7 ± 6.3
Macrohardness (HR15T)	37 ± 1	—	—	65 ± 1
Microhardness (HV)—transverse^a	40.0 ± 0.2	51.0 ± 0.7	56.3 ± 0.7	65.9 ± 0.9
Microhardness (HV)—longitudinal^a	40.1 ± 0.3	50.1 ± 0.7	54.5 ± 1.3	66.2 ± 0.6

^{a}Matrix microhardness values.

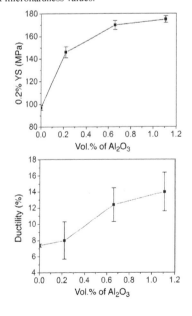

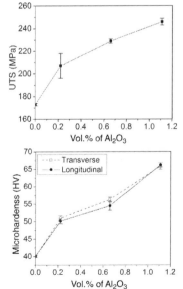

5.3.2.2. Mg Reinforced with 50 nm Al₂O₃ (by Powder Metallurgy—Conventional Sintering)

Processing method	Powder metallurgy route
Sintering method	Conventional tube furnace sintering
Matrix material used	Pure Mg (99.5% purity)
Size of Al_2O_3 used	50 nm
Extrusion temperature	200°C

Refer to Table 5.10.

TABLE 5.10. Characteristics of Mg and Mg/Al₂O₃ [21].

Material	Mg	Mg/1 Al₂O₃	Mg/3 Al₂O₃	Mg/5 Al₂O₃
Vol.% of Al₂O₃	—	1.0	3.0	5.0
CTE ($\times 10^{-6}$/°C)	30.6 ± 0.2	28.9 ± 0.3	26.9 ± 0.1	22.0 ± 0.0
0.2% YS (MPa)	133 ± 12	125 ± 15	118 ± 2	102 ± 2
UTS (MPa)	172 ± 21	158 ± 19	148 ± 8	136 ± 5
Microhardness (HV)—matrix	37.9 ± 0.9	43.7 ± 1.5	47.0 ± 3.4	48.4 ± 3.8

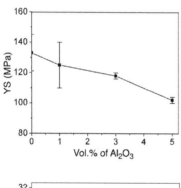

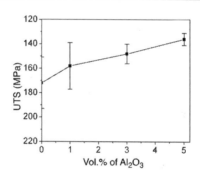

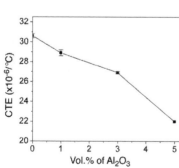

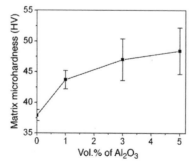

5.3.2.3. Mg Reinforced with 50 nm Al₂O₃ (by Powder Metallurgy—Microwave Sintering)

Processing method	Powder metallurgy route
Sintering method	Hybrid microwave-assisted sintering
Matrix material used	Pure Mg (98.5% purity)
Size of Mg used	60–300 μm
Size of Al₂O₃ used	50 nm
Sintering time	25 min
Extrusion temperature	350°C
Extrusion ratio	25:1

Refer to Table 5.11.

TABLE 5.11. Characteristics of Mg and Mg/Al₂O₃ [20].

Material	Mg	Mg/0.3 Al₂O₃	Mg/0.6 Al₂O₃	Mg/1.0 Al₂O₃	Mg/1.5 Al₂O₃
Wt% of Al₂O₃ (vol.%)	—	0.7 (0.3)	1.4 (0.6)	2.2 (1.0)	3.3 (1.5)
Density (g/cm³)	1.738	1.741	1.742	1.748	1.756
Porosity (%)	0.12	0.32	0.67	0.83	1.00
0.2% YS (MPa)	116 ± 11	119 ± 7	130 ± 5	154 ± 5	148 ± 10
UTS (MPa)	168 ± 10	175 ± 8	180 ± 7	213 ± 12	209 + 7
Failure strain (%)	6.1 ± 2.0	7.5 ± 0.2	7.4 ± 0.3	6.3 ± 0.4	5.6 ± 0.3
Work of fracture (MJ/m³)	11.8 ± 3.4	12.8 ± 0.9	13.4 ± 0.5	13.2 ± 0.9	11.5 ± 0.8
Microhardness (HV)	40 ± 1	48 ± 3	54 ± 3	60 ± 4	68 ± 2

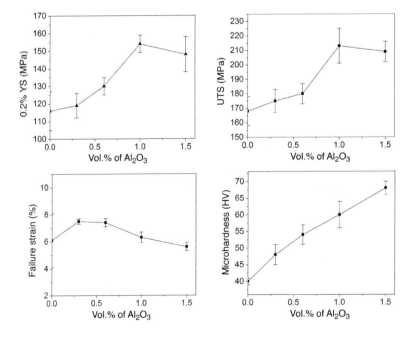

5.3.2.4. AZ31B Reinforced with 50 nm Al₂O₃ (by Disintegrated Melt Deposition)

Processing method	Disintegrated melt deposition
Matrix material used	AZ31B (2.94% Al, 0.87% Zn, 0.57% Mn, 0.0027% Fe, 0.0112% Si, 0.0008% Cu, 0.0005% Ni, and balance Mg)
Size of Al₂O₃ used	50 nm
Extrusion temperature	350°C
Extrusion ratio	20.25:1

Refer to Table 5.12.

TABLE 5.12. Characteristics of AZ31B and AZ31B/Al₂O₃ composites [13].

Material	AZ31B	AZ31B/0.66 Al₂O₃	AZ31B/1.11 Al₂O₃	AZ31B/1.50 Al₂O₃
Vol.% of Al₂O₃	—	0.66	1.11	1.50
Density (g/cm³)	1.770	1.784	1.793	1.809
Porosity (%)	0	0.06	0.06	0.07
CTE	26.2	25.7	25.1	24.7
0.2% YS (MPa)	201 ± 7	149 ± 7	148 ± 11	144 ± 9
UTS (MPa)	270 ± 6	215 ± 15	214 ± 16	214 ± 16
Failure strain (%)	5.6 ± 1.4	14.6 ± 1.1	25.5 ± 2.2	29.5 ± 1.9
Work of fracture (J/m³)	15 ± 3	31 ± 4	52 ± 2	60 ± 3
Microhardness (HV)—matrix	63 ± 1	79 ± 4	82 ± 2	86 ± 3

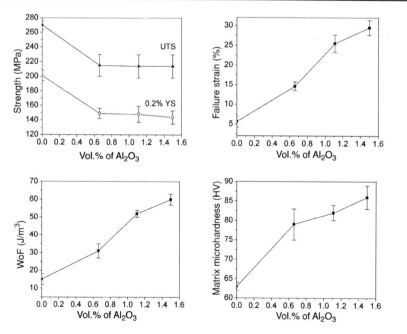

5.3.2.5. AZ31B Reinforced with 50 nm Al₂O₃ and with Ca addition (by Disintegrated Melt Deposition)

Processing method   Disintegrated melt deposition
Matrix material used   AZ31B (2.94% Al, 0.87% Zn, 0.57% Mn,
                           0.0027% Fe, 0.0112% Si, 0.0008% Cu,
                           0.0005% Ni, and balance Mg)
Size of Al₂O₃ used   50 nm
Size of Ca used   6 mesh
Extrusion temperature   350°C
Extrusion ratio   20.25:1

Refer to Table 5.13.

TABLE 5.13. Characteristics of AZ31B and AZ31B-based composites [14].

Material	(1) AZ31B	(2) AZ31B/3.3 Al₂O₃	(3) AZ31B/3.3 Al₂O₃–1Ca	(4) AZ31B/3.3 Al₂O₃–2Ca	(5) AZ31B/3.3 Al₂O₃–3Ca
Wt% of Al₂O₃ (vol.%)	—	3.3 (1.5)	3.3 (1.5)	3.3 (1.5)	3.3 (1.5)
Wt% of Ca	—	—	1	2	3
Density (g/cm³)	1.7679	1.8086	1.7932	1.7956	1.7922
Porosity (%)	0.12	0.07	0.36	0.08	0.10
0.2% YS (MPa)	201 ± 7	144 ± 9	185 ± 6	215 ± 5	235 ± 7
UTS (MPa)	270 ± 6	214 ± 16	243 ± 17	261 ± 8	285 ± 14
Failure strain (%)	5.6 ± 1.4	29.5 ± 1.9	16 ± 1.2	10 ± 0.2	7.3 ± 0.2
Microhardness (HV)—matrix	63 ± 1	86 ± 3	98 ± 3	104 ± 3	113 ± 4

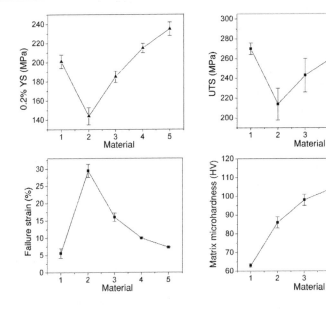

5.3.2.6. AZ31 Reinforced with 50 nm Al₂O₃ (by Disintegrated Melt Deposition)

Processing method	Disintegrated melt deposition
Matrix material used	AZ31 (2.50–3.50 wt% Al, 0.60–1.40 wt% Zn, 0.15–0.40 wt% Mn, 0.10 wt% Si, 0.05 wt% Cu, 0.01 wt% Fe, 0.01 wt% Ni, and balance Mg)
Size of Al₂O₃ used	50 nm
Extrusion temperature	350°C
Extrusion ratio	20.25:1

Refer to Tables 5.14 and 5.15.

TABLE 5.14. Results of grain characteristics, microhardness and tensile testing of AZ31 and AZ31/1.5 vol.% Al₂O₃ composite [12].

Material	AZ31	AZ31B/1.5 Al₂O₃
Vol.% of Al₂O₃	—	1.5
Grain size (μm)	4.0 ± 0.9	2.3 ± 0.7
Grain aspect ratio	1.4	1.6
Microhardness (HV)	64 ± 4	83 ± 5
0.2% TYS (MPa)	172 ± 15	204 ± 8
UTS (MPa)	263 ± 12	317 ± 5
Failure strain (%)—tensile	10.4 ± 3.9	22.2 ± 2.4
WOF (MJ/m³)—tensile	26 ± 9	68 ± 7

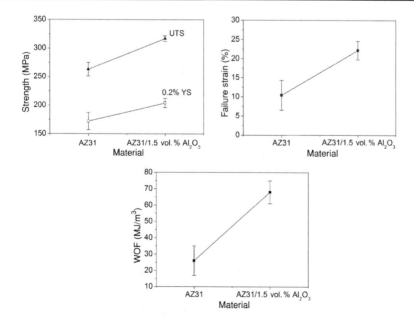

TABLE 5.15. Compressive test results of AZ31 and AZ31/1.5 vol.% Al₂O₃ composite [12].

Material	AZ31	AZ31B/1.5Al₂O₃
0.2% CYS (MPa)	93 ± 9	98 ± 2
UCS (MPa)	486 ± 4	509 ± 12
Failure strain (%)—compressive	19.7 ± 7.2	19.0 ± 2.7
WOF (MJ/m³)—compressive	76 ± 14	84 ± 15

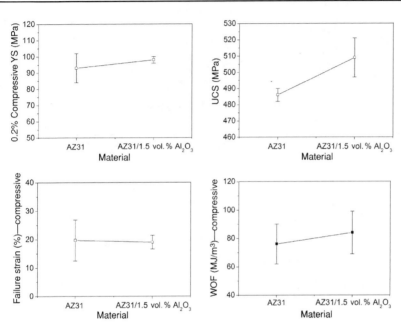

5.3.3. Addition of Hybrid Reinforcements (with Al₂O₃)

5.3.3.1. Mg Reinforced with Al₂O₃ of Different Sizes (by Powder Metallurgy—Conventional Sintering)

Processing method	Powder metallurgy route
Sintering method	Conventional tube furnace sintering
Matrix material used	Pure Mg ($\geq 98.5\%$ purity)
Size of Mg used	60–300 μm
Size of Al₂O₃ used	50 nm, 0.3 μm, 1.0 μm
Amount% of Al₂O₃	1.1 vol.% (2.5 wt%)
Extrusion ratio	20.25:1
Extrusion temperature	250°C

Refer to Table 5.16.

TABLE 5.16. Characteristics of Mg and Mg/Al$_2$O$_3$ [8, 10].

Material	(1) Mg	(2) Mg/Al$_2$O$_3$ (50 nm)	(3) Mg/Al$_2$O$_3$ (0.3 μm)	(4) Mg/Al$_2$O$_3$ (1.0 μm)
Density (g/cm^3)	1.7387	1.7632	1.7646	1.7645
Porosity (%)	0.08	0.09	0.01	0.01
0.2% YS (MPa)	132 ± 7	194 ± 5	182 ± 3	172 ± 1
UTS (MPa)	193 ± 2	250 ± 3	237 ± 1	227 ± 2
Ductility (%)	4.2 ± 0.1	6.9 ± 1.0	12.1 ± 1.4	16.8 ± 0.4
Macrohardness (HR15T)	43.5 ± 0.3	59.7 ± 0.5	56.3 ± 0.5	50.3 ± 0.5
Microhardness (HV)—matrix	37.4 ± 0.4	69.5 ± 0.5	51.8 ± 0.3	51.2 ± 0.5

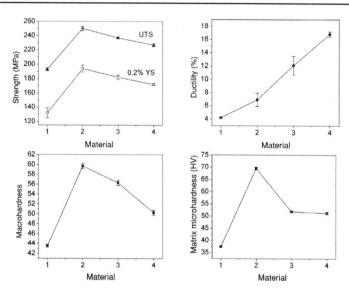

5.3.3.2. Mg Reinforced with Al$_2$O$_3$ of Different Size (by Powder Metallurgy—Microwave Sintering)

Processing method	Powder metallurgy route
Sintering method	Hybrid microwave-assisted sintering
Matrix material used	Pure Mg (98.5% purity)
Size of Mg used	60–300 μm
Size of Al$_2$O$_3$ used	50 nm, 0.3 μm
Sintering time	25 min
Extrusion temperature	350°C
Extrusion ratio	25:1

Refer to Table 5.17.

TABLE 5.17. Characteristics of Mg and Mg/Al$_2$O$_3$ [15].

Material	(1) Mg	(2) Mg/Al$_2$O$_3$	(3) Mg/Al$_2$O$_3$	(4) Mg/Al$_2$O$_3$
Vol.% of Al$_2$O$_3$	—	0.50% (50 nm) 4.50% (0.3 μm)	0.75% (50 nm) 4.25% (0.3 μm)	1.00% (50 nm) 4.00% (0.3 μm)
Density (g/cm^3)	1.738	1.840	1.837	1.831
Porosity (%)	0.07	0.58	0.75	1.04
CTE ($\times 10^{-6}$/°C)	28.6 $\pm$ 0.8	27.2 $\pm$ 1.2	25.7 $\pm$ 0.6	25.8 $\pm$ 0.8
Elastic modulus (GPa)	45	50.5	51.9	54.4
0.2% YS (MPa)	116 $\pm$ 11	139 $\pm$ 27	138 $\pm$ 13	157 $\pm$ 20
UTS (MPa)	168 $\pm$ 10	187 $\pm$ 28	189 $\pm$ 15	211 $\pm$ 21
Ductility (%)	9.0 $\pm$ 0.3	1.9 $\pm$ 0.2	2.4 $\pm$ 0.6	3.0 $\pm$ 0.3
Microhardness (HV)—matrix	47.0 $\pm$ 1.3	56.6 $\pm$ 1.2	86.7 $\pm$ 1.7	73.7 $\pm$ 1.1

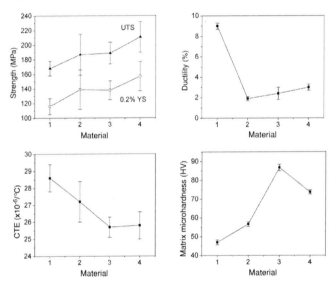

5.3.3.3. Mg Reinforced with Mg-NanoAl$_2$O$_3$ Concentric Alternating Macro Ring (by Powder Metallurgy—Microwave Sintering)

Processing method	Powder metallurgy route
Sintering method	Hybrid microwave-assisted sintering
Matrix material used	Pure Mg (98.5% purity)
Size of Mg used	60–300 μm
Size of Al$_2$O$_3$ used	50 nm
Amount of Al$_2$O$_3$ used	1.11 vol.%
Sintering time	34 min
Extrusion temperature	350°C
Extrusion ratio	20.25:1

Refer to Table 5.18.

TABLE 5.18. Characteristics of Mg and Mg-Mg/1.11 vol.% Al_2O_3 ring structured hybrid composites [11].

Material	Mg	Mg-Mg/1.11 vol.% Al_2O_3			
	(1)	(2)	(3)	(4)	(5)
Thickness of alternating layers	—	4 mm	3 mm	2 mm	1 mm
0.2% YS (MPa)	115 ± 3	125 ± 3	149 ± 5	112 ± 2	102 ± 5
UTS (MPa)	195 ± 11	195 ± 6	223 ± 10	173 ± 9	151 ± 7
Failure strain (%)	11 ± 3	8 ± 1	13 ± 1	6 ± 3	6 ± 2

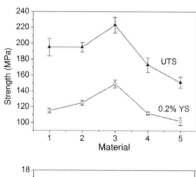

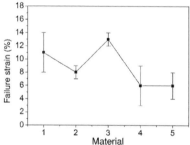

5.3.3.4. Mg Reinforced with Al₂O₃ and MWCNT (by Powder Metallurgy—Microwave Sintering)

Processing method	Powder metallurgy route
Sintering method	Hybrid microwave-assisted sintering
Matrix material used	Pure Mg (98.5% purity)
Size of Mg used	60–300 μm
Size of Al₂O₃ used	50 nm
Multiwalled CNT used	40–70 nm (outer diameter)
Sintering time	25 min
Extrusion temperature	350°C
Extrusion ratio	25:1

Refer to Table 5.19.

TABLE 5.19. Characteristics of Mg and Mg-based composites [16].

Material	(1) Mg 1 CNT	(2) Mg 0.7 CNT 0.3 Al_2O_3	(3) Mg 0.5 CNT 0.5 Al_2O_3	(4) Mg 0.3 CNT 0.7 Al_2O_3
Wt% of reinforcement (type)	1.0% (MWCNT)	0.7% (MWCNT) 0.3% (Al_2O_3)	0.5% (MWCNT) 0.5% (Al_2O_3)	0.3% (MWCNT) 0.7% (Al_2O_3)
Density (g/cm^3)	1.735	1.737	1.745	1.730
Porosity (%)	0.61	0.59	0.19	1.07
0.2% YS (MPa)	113 ± 3	131 ± 6	137 ± 6	154 ± 2
UTS (MPa)	147 ± 7	164 ± 11	181 ± 9	196 ± 3
Failure strain (%)	1.9 ± 0.9	2.6 ± 1.3	2.5 ± 0.4	2.5 ± 0.8
Macrohardness (HR15T)	47.9 ± 2.6	48.0 ± 0.8	48.2 ± 1.5	48.3 ± 1.5
Microhardness (HV)—matrix	43.2 ± 1.6	43.5 ± 1.4	43.7 ± 2.3	44.2 ± 1.8

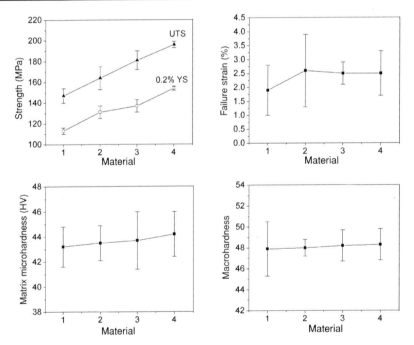

5.4. MAGNESIUM-BASED COMPOSITES WITH MgO

This section investigates the effect of presence of MgO in nanolength scale on the properties of pure magnesium. Results reveal that the best combination of tensile properties (in terms of strength and ductility) can be realized with the addition of 0.5 vol.% of MgO.

5.4.1. Addition of Nanosize MgO

5.4.1.1. Mg Reinforced with 36 nm MgO (by Disintegrated Melt Deposition)

Processing method	Disintegrated melt deposition
Matrix material used	Pure Mg (>99.9% purity)
Size of MgO used	36 nm
Extrusion temperature	350°C
Extrusion ratio	20.25:1

Refer to Table 5.20.

TABLE 5.20. Characteristics of Mg and Mg/MgO [29].

Material	Mg	Mg/0.5 MgO	Mg/0.75 MgO	Mg/1.0 MgO
Vol.% of MgO	—	0.5	0.75	1.0
Density (g/cm^3)	1.738	1.748	1.751	1.755
CTE ($\times 10^{-6}$/°C)	28.7 ± 0.6	26.9 ± 0.7	26.0 ± 0.7	25.6 ± 0.1
Elastic modulus (GPa)	40 ± 2	49 ± 1	54 ± 1	54 ± 2
0.2% YS (MPa)	126 ± 7	151 ± 3	158 ± 5	169 ± 8
UTS (MPa)	192 ± 5	233 ± 5	213 ± 4	223 ± 8
Ductility (%)	8 ± 2	8 ± 1	3 ± 2	3 ± 1
Macrohardness (HR15T)	45 ± 1	47 ± 1	53 ± 1	54 ± 2

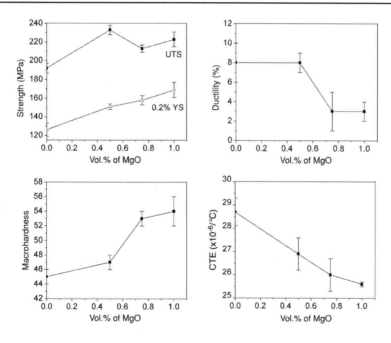

5.5. MAGNESIUM-BASED COMPOSITES WITH SiC

SiC is one of the most investigated reinforcement in magnesium and its alloys. Most commonly used magnesium matrices include pure magnesium, Mg-Zn, and AZ91 alloys. Most commonly investigated length scale is micrometer length scale, while limited studies have been done on sub-micrometer and nanolength scales of SiC. Processing methods such as stir casting, liquid filtration, disintegrated melt deposition, vacuum stir casting, ultrasonic cavitation processing, and powder metallurgy techniques involving solid-phase sintering, liquid-phase sintering, and microwave sintering are investigated. In addition, effects of extrusion temperature, heat treatment, and recycling on the end properties of the base matrix are also investigated.

5.5.1. Addition of Micrometer-Size SiC

5.5.1.1. Mg and AZ91D Reinforced with 150 μm SiC (by Stir Casting)

Processing method	Stir casting process
Matrix materials used	Mg (99.8% purity) and AZ91D
Reinforcement used	α-SiC
Size of SiC used	Average 150 μm

Refer to Table 5.21.

TABLE 5.21. Characteristics of monolithic Mg, AZ91D alloy, and composites [30].

Material	Mg	Mg/15 SiC	AZ91D	AZ91D/15 SiC
Wt% of SiC (vol.%)	—	24 (15)	—	24 (15)
Density (g/cm³)	1.738	1.934	1.807	1.990
Porosity (vol.%)	0.11	1.33	0.14	1.37
Macrohardness (±2 HV)	44.9	90.3	65.7	91.2
Microhardness (HV)— matrix	51.4 ± 0.5	80.6 ± 0.6	77.0 ± 0.9	81.4 ± 1.1
Microhardness (HV)— matrix/SiC interface	—	113.6 ± 6.0	—	139.4 ± 4.0

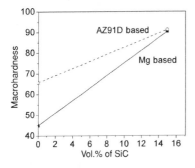
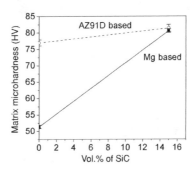

5.5.1.2. Mg and AZ91 Reinforced with 100µm SiC (by Liquid Infiltration)

Processing method   Liquid infiltration process
Matrix materials used   Pure Mg and AZ91
Size of Mg and AZ91 used   100–150 μm (Approx.)
Size of SiC used   Average 100 μm

Refer to Table 5.22.

TABLE 5.22. Characteristics of Mg/SiC and AZ91/SiC composites [31].

Material	Mg/20 SiC	AZ91/20 SiC	Mg/30 SiC	AZ91/30 SiC	Mg/40 SiC	AZ91/40 SiC
Vol.% of SiC	20	20	30	30	40	40
Microhardness (HVN)	6315	8109	7258	9138	8865	10594

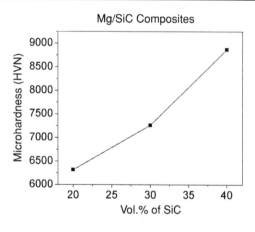

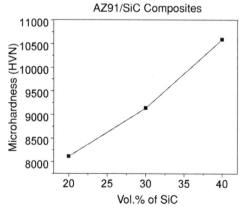

5.5.1.3. Mg Reinforced with 40 μm SiC (by Melt Stir Technique)

Processing method   Melt stir technique
Matrix material used   Pure Mg (99% purity)
Size of SiC used   Average 40 μm
Extrusion ratio   13:1 (prior to hot extrusion, billets were
                         homogenized at 500 for 2 h)

(a) Tested at Room Temperature
Refer to Table 5.23.

TABLE 5.23. Characteristics of Mg and Mg/SiC composite at room temperature [32].

Material	Mg	Mg/30 SiC
Vol.% of SiC	—	30
Young's modulus (GPa)	38	59
0.2% Proof stress (MPa)	135	229
UTS (MPa)	196	258
Elongation (%)	12	2
Macrohardness (VHN)	45	60
Microhardness (VHN)	43	55

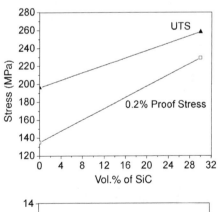

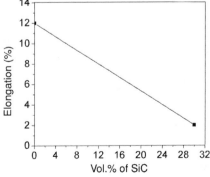

(b) Tested at Elevated Temperatures
Refer to Table 5.24.

TABLE 5.24. Tensile properties of Mg and Mg/SiC composites at elevated temperatures [32].

Material	Mg-250T	Mg/SiC-250T	Mg-300T	Mg/SiC-300T	Mg-350T	Mg/SiC-350T
Vol.% of SiC	—	30	—	30	—	30
Testing temperature (°C)	250	250	300	300	350	350
0.2% Proof stress (MPa)	47	68	33	42	15	31
UTS (MPa)	68	91	46	57	22	42
Elongation (%)	48	20	68	22	43	16

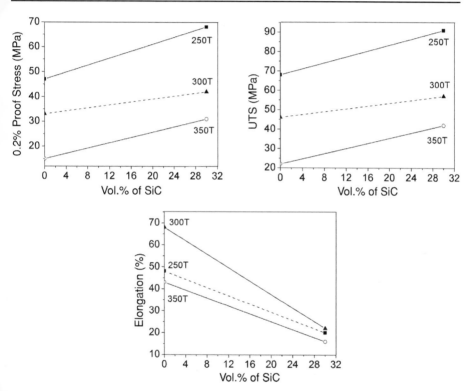

5.5.1.4. Mg Reinforced with 38 μm SiC (by Disintegrated Melt Deposition)

Processing method	Disintegrated melt deposition
Matrix material used	Pure Mg (99.5% purity)
Size of SiC used	38 μm
Extrusion ratio	13:1

(a) Effect of Extrusion Temperature

Extrusion temperature   350°C, 250°C, 150°C, and 100°C

Refer to Table 5.25.

TABLE 5.25. Characteristics of Mg/SiC subjected to different extrusion temperatures [33].

Material	Mg/SiC-350E	Mg/SiC-250E	Mg/SiC-150E	Mg/SiC-100E
Wt% of SiC (vol.%)	11.5 (6.6)	11.5 (6.6)	11.5 (6.6)	11.4 (6.5)
Density (g/cm^3)	1.829	1.828	1.827	1.826
Porosity (%)	0.48	0.50	0.56	0.56
Elastic modulus (GPa)	42.7 ± 0.1	43.3 ± 0.3	49.5 ± 1.9	50.8 ± 1.4
0.2% YS (MPa)	119 ± 3	139 ± 7	148 ± 3	148 ± 7
UTS (MPa)	197 ± 3	201 ± 12	204 ± 8	201 ± 7
Ductility (%)	3.9 ± 0.2	3.7 ± 0.7	3.6 ± 0.9	3.6 ± 1.1
Macrohardness (HR15T)	52.5 ± 0.9	54.8 ± 0.7	57.1 ± 0.6	57.7 ± 0.6
Microhardness (HV)—matrix	42.7 ± 0.6	44.4 ± 0.6	45.3 ± 0.7	45.7 ± 0.7
Microhardness (HV)—Mg/SiC interface	135 ± 16	177 ± 24	189 ± 34	253 ± 35

350E, extrusion at 350°C; 250E, extrusion at 250°C; 150E, extrusion at 150°C; 100E, extrusion at 100°C.

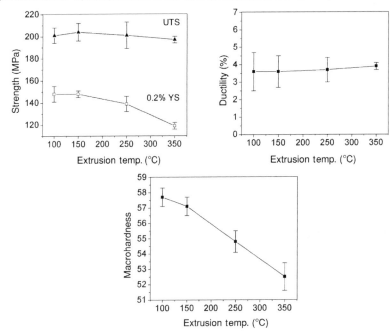

(b) Effect of Heat Treatment

Extrusion at   100°C
Heat treated at 100°C   5 h and 10 h

Refer to Table 5.26.

TABLE 5.26. Characteristics of Mg/SiC subjected to different heat treatments [33, 34].

Material	Mg/SiC-HT0	Mg/SiC-HT5	Mg/SiC-HT10
Wt% of SiC (vol.%)	11.4 (6.5)	11.4 (6.5)	11.4 (6.5)
Elastic modulus (GPa)	50.8 ± 1.4	50.0 ± 2.5	50.8 ± 1.7
0.2% YS (MPa)	148 ± 7	130 ± 12	126 ± 8
UTS (MPa)	201 ± 7	191 ± 11	190 ± 8
Ductility (%)	3.6 ± 1.1	12.8 ± 1.5	10.4 ± 1.6
Macrohardness (HR15T)	57.7 ± 0.6	57.4 ± 0.4	57.3 ± 0.3
Microhardness (HV)—matrix	45.7 ± 0.7	45.4 ± 0.6	44.5 ± 0.7
Microhardness (HV)–Mg/SiC interface	253 ± 35	217 ± 23	189 ± 12

HT, not heat treated; HT5, heat treated for 5 h; HT10, heat treated for 10 h.

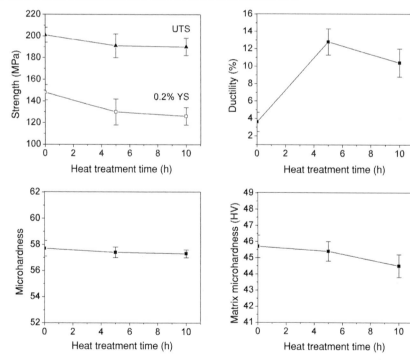

(c) Effect of Recycling

Processing method	Disintegrated melt deposition
Matrix material used	Pure Mg (99.9% purity)
Size of SiC used	38 μm
Extrusion temperature	350°C
Extrusion ratio	13:1

Refer to Table 5.27.

TABLE 5.27. Characteristics of Mg/SiC subjected to recycling [35].

Material	Mg/SiC-N	Mg/SiC-R1	Mg/SiC-R2
Wt% of SiC (vol.%)	12.7 (7.3)	11.5 (6.6)	11.0 (6.3)
Density (g/cm^3)	1.828	1.822	1.829
Porosity (%)	1.11	0.80	0.21
CTE ($\times 10^{-6}/$°C)	25.7 ± 0.2	25.7 ± 0.2	26.1 ± 0.1
Elastic modulus (GPa)	43 ± 2	43 ± 2	44 ± 3
0.2% YS (MPa)	116 ± 1	122 ± 7	126 ± 10
UTS (MPa)	176 ± 4	195 ± 2	200 ± 3
Ductility (%)	3.0 ± 0.5	4.7 ± 0.4	4.4 ± 0.0
Macrohardness (HR15T)	50.4 ± 1.0	50.0 ± 0.7	49.9 ± 0.8
Microhardness (HV)—matrix	44 ± 1	41 ± 1	39 ± 1
Microhardness (HV)–Mg/SiC interface	116 ± 14	113 ± 10	108 ± 9

N, new; R1, recycled once, R2, recycled twice.

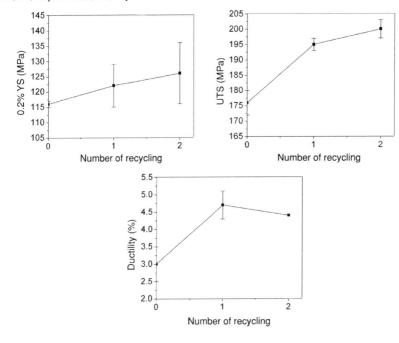

5.5.1.5. Mg Reinforced with 35 μm SiC (by Conventional Casting)

Processing method   Conventional casting
Matrix material used   Pure Mg (99.5% purity)
Size of SiC used   35 μm
Extrusion temperature   350°C
Extrusion ratio   13:1

Refer to Table 5.28.

TABLE 5.28. Characteristics of Mg and Mg/SiC [36].

Material	Mg	Mg/33.6 SiC
Wt% of SiC	—	33.6
Density (g/cm^3)	1.73	2.03
Porosity (%)	0.58	1.35
0.2% YS (MPa)	153 ± 3	158 ± 8
UTS (MPa)	219 ± 1	205 ± 6
Ductility (%)	12.1 ± 1.4	3.1 ± 0.3
Macrohardness (HR15T)	42.6 ± 0.5	62.2 ± 0.5

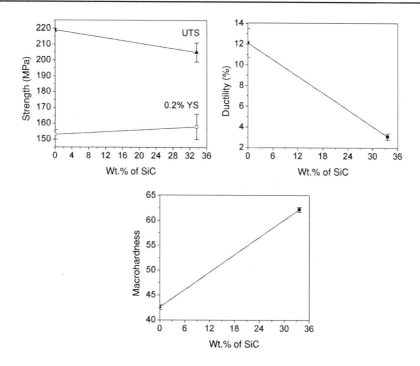

5.5.1.6. Mg Reinforced with 25 μm SiC (by Disintegrated Melt Deposition)

Processing method	Disintegrated melt deposition
Matrix material used	Pure Mg (99.5% purity)
Size of SiC used	25 μm
Extrusion temperature	350°C
Extrusion ratio	13:1

Refer to Table 5.29.

TABLE 5.29. Characteristics of Mg and Mg/SiC [37, 38].

Material	Mg	Mg/5.8SiC	Mg/9.3SiC	Mg/12.8SiC
Wt% of SiC (vol.%)	—	10.3 (5.8)	16.0 (9.3)	21.3 (12.8)
Density (g/cm^3)	1.73	1.82	1.85	1.92
Porosity (%)	0.82	0.12	1.24	0.68
CTE ($\times 10^{-6}$ /°C)	28.43	25.36	23.65	23.56
0.2% YS (MPa)	126 ± 14	127 ± 7	120 ± 5	128 ± 2
UTS (MPa)	200 ± 5	195 ± 7	181 ± 6	176 ± 4
Ductility (%)	11.7 ± 3.1	6.0 ± 2.3	4.7 ± 1.3	1.4 ± 0.1
Macrohardness (HR15T)	75.1 ± 1.7	79.0 ± 1.1	80.1 ± 1.1	82.2 ± 1.1
Microhardness (HV)—matrix	39.7 ± 1.6	41.2 ± 2.9	42.9 ± 5.8	43.0 ± 1.0
Microhardness (HV)—Mg/SiC interface	—	217.9 ± 17.3	224.8 ± 19.8	261.7 ± 27.6

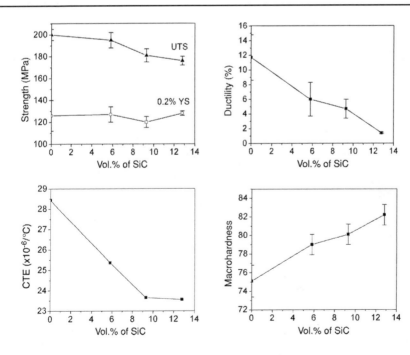

5.5.1.7. Mg Reinforced with 25 μm SiC (by Conventional Casting)

Processing method   Conventional casting
Matrix material used   Pure Mg (99.5% purity)
Size of SiC used   25 μm
Extrusion temperature   350°C
Extrusion ratio   13:1

Refer to Table 5.30.

TABLE 5.30. Characteristics of Mg and Mg/SiC [39].

Material	Mg	Mg/4.3 SiC	Mg/8.7 SiC	Mg/16.0 SiC
Wt% of SiC (vol.%)	—	7.6 (4.3)	14.9 (8.7)	26.0 (16.0)
Density (g/cm^3)	1.739	1.788	1.847	1.958
Porosity (%)	0.06	0.84	1.14	0.92
CTE ($\times 10^{-6}$/°C)	29.35	27.16	24.33	23.38
Dynamic elastic modulus (GPa)	42.6	45.0	49.4	54.1
0.2% YS (MPa)	110 ± 2	112 ± 6	112 ± 2	114 ± 4
UTS (MPa)	199 ± 1	191 ± 2	177 ± 4	154 ± 8
Ductility (%)	10.2 ± 0.3	6.8 ± 0.1	4.7 ± 0.3	1.8 ± 0.7
Macrohardness (HR15T)	40.0 ± 0.6	45.3 ± 0.8	49.5 ± 0.8	57.6 ± 1.2
Microhardness (HV)—matrix	47.2 ± 0.4	48.9 ± 0.6	49.1 ± 1.5	49.9 ± 1.1
Microhardness (HV)–Mg/SiC interface	—	99 ± 18	102 ± 11	113 ± 16

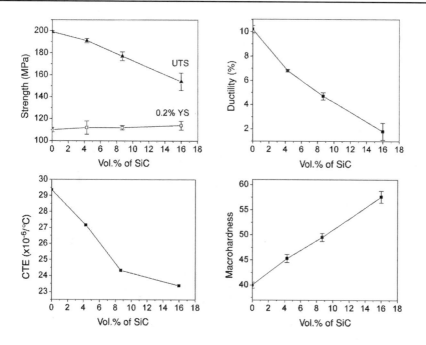

5.5.1.8. Mg–Al Reinforced with 20 μm SiC (by Powder Metallurgy)

Processing method	Powder metallurgy route
Matrix materials used	Pure Mg (>97.5% purity) and pure Al (>96% purity)
Size of Mg used	Average 50 μm
Size of Al used	Average 13.3 μm
Size of SiC used	Average 20 μm
Extrusion temperature	420°C
Extrusion ratio	10:1

(a) Liquid-Phase Sintering
  •470°C for 1 h
Refer to Table 5.31.

TABLE 5.31. Characteristics of Mg–Al alloy and Mg–Al/SiC composite [40].

Material	Mg–9 wt% Al	Mg–9 wt% Al/15 vol.% SiC
Vol.% of SiC	—	15
Grain size (μm)	6.6	3.3
0.2% Yield strength (MPa)	276	339
Tensile strength (MPa)	327	364
Elongation (%)	5.8	1.2

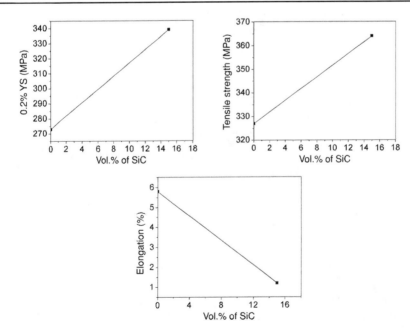

(b) Solid-Phase Sintering
 • 427°C for 10 h
Refer to Table 5.32.

TABLE 5.32. Characteristics of Mg–Al alloy and Mg–Al/SiC composite [40].

Material	Mg–9 wt% Al	Mg–9 wt% Al/15 vol.% SiC
Vol.% of SiC	—	15
Grain size (μm)	12.0	7.3
0.2% Yield strength (MPa)	239	271
Tensile strength (MPa)	345	355
Elongation (%)	8.0	3.2

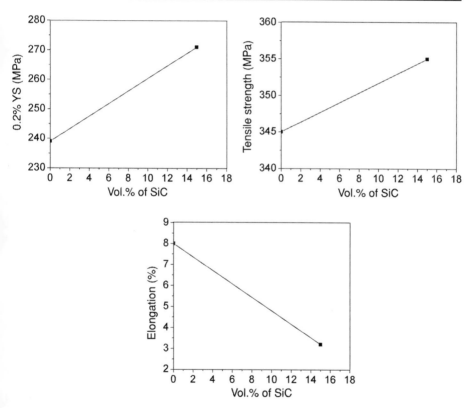

5.5.1.9. AZ91C Reinforced with 12.8 μm SiC (by Vacuum Stir Casting)

Processing method	Vacuum stir casting process
Matrix materials used	Mg–Al9Zn alloy (AZ91C)
Size of SiC used	Average 12.8 μm
Heat treatment	T4 (18 h at 415°C under CO_2 atmosphere)

Refer to Table 5.33.

TABLE 5.33. Tensile properties of monolithic Mg–Al9Zn alloy and Mg–Al9Zn/SiC composite [41].

Material	Mg–Al9Zn	Mg–Al9Zn/15 SiC
Vol.% of SiC	—	15
Elastic modulus (GPa)	42.7	57.0
Yield strength (MPa)	84	178
Tensile strength (MPa)	225	218
Elongation (%)	7.2	1.1

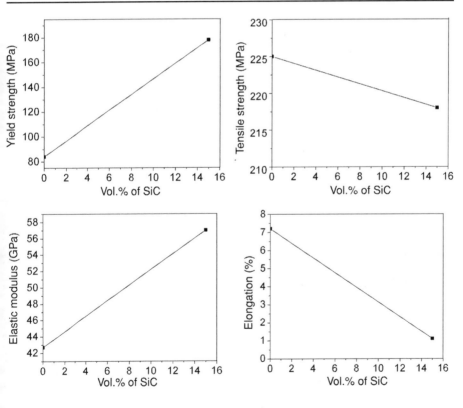

5.5.2. Addition of Sub-Micrometer-Size SiC

5.5.2.1. Mg Reinforced with 0.6 μm SiC (by Disintegrated Melt Deposition)

Processing method	Disintegrated Melt Deposition
Matrix material used	Pure Mg (99.9% purity)
Size of SiC used	0.6 μm
Extrusion temperature	350°C
Extrusion ratio	20.25:1

Refer to Table 5.34.

TABLE 5.34. Characteristics of Mg and Mg/SiC [42, 43].

Material	Mg	Mg/2.7 SiC	Mg/5.8 SiC	Mg/9.0 SiC
Wt% of SiC (vol.%)	—	4.8 (2.7)	10.2 (5.8)	15.4 (9.0)
Density (g/cm^3)	1.7380	1.7698	1.7931	1.8349
Porosity (%)	0.12	0.53	1.75	1.98
CTE ($\times 10^{-6}$ /°C)	28.02	23.77	21.20	21.04
Dynamic elastic modulus (GPa)	39.8	45.6	47.2	48.2
0.2% YS (MPa)	153 ± 8	182 ± 2	171 ± 3	155 ± 1
UTS (MPa)	207 ± 4	219 ± 2	221 ± 14	207 ± 9
Ductility (%)	9.2 ± 1.4	2.1 ± 0.9	1.5 ± 0.2	1.4 ± 0.1
Macrohardness (HR15T)	47 ± 1	58 ± 1	59 ± 1	63 ± 1
Microhardness (HV)—matrix	41 ± 1	53 ± 1	55 ± 1	56 ± 2

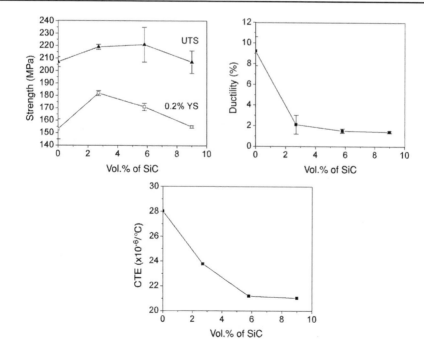

Effect of Heat Treatment
- Heat treated at 150°C for 5 h

Refer to Table 5.35.

TABLE 5.35. Characteristics of not heat-treated Mg/SiC and heat-treated Mg/SiC [42, 43].

Material	Mg/2.7 SiC-HT0	Mg/9 SiC-HT0	Mg/2.7 Si-HT5	Mg/9 Si-HT5
Wt% of SiC (vol.%)	4.8 (2.7)	15.4 (9.0)	4.8 (2.7)	15.4 (9.0)
CTE ($\times 10^{-6}$ /°C)	23.77	21.04	23.85	22.21
0.2% YS (MPa)	182 ± 2	155 ± 1	200 ± 9	168 ± 5
UTS (MPa)	219 ± 2	207 ± 9	233 ± 14	213 ± 4
Ductility (%)	2.1 ± 0.9	1.4 ± 0.1	2.9 ± 0.3	3.6 ± 0.9
Macrohardness (HR15T)	58 ± 1	63 ± 1	56 ± 2	60 ± 1
Microhardness (HV—matrix	53 ± 1	56 ± 2	51 ± 2	54 ± 3

HT0, not heat treated; HT5, heat treated at 150°C for 5 h.

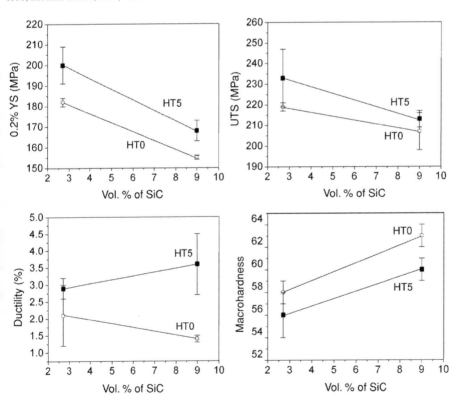

5.5.3. Addition of Nanosize SiC

5.5.3.1. Mg Reinforced with 50 nm SiC (by Melt Casting—Ultrasonic Cavitation)

Processing method	Melt casting (ultrasonic cavitation processing)
Matrix materials used	Pure Mg
Size of SiC used	Average 50 nm

Refer to Table 5.36.

TABLE 5.36. Characteristics of Mg and Mg/SiC composites [44].

Material	Mg	Mg/0.5% SiC	Mg/1% SiC	Mg/2% SiC
Amount of SiC (wt%)	—	0.5	1.0	2.0
Yield strength (MPa)	20.0	28.3	30.3	35.9
UTS (MPa)	89.6	120.7	124.1	131.0
Ductility (%)	14.0	15.5	14.2	12.6

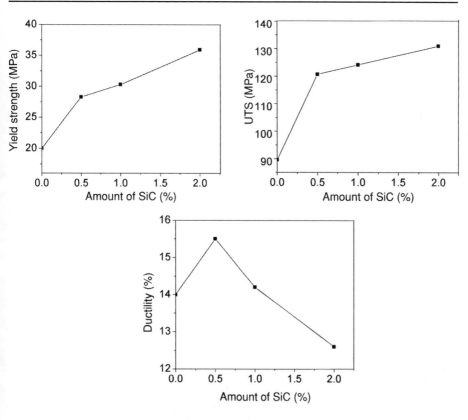

5.5.3.2. Mg4Zn Reinforced with 50 nm SiC (by Melt Casting—Ultrasonic Cavitation)

Processing method   Melt casting (ultrasonic cavitation processing)
Matrix materials used   Mg–4Zn alloy
Size of SiC used   Average 50 nm

Refer to Table 5.37.

TABLE 5.37. Characteristics of Mg4Zn and Mg4Zn/SiC composites [45].

Material	Mg4Zn	Mg4Zn/1.5% SiC
Amount of SiC (wt%)	—	1.5
Yield strength (MPa)	42	72
Tensile strength (MPa)	105	199
Ductility (%)	8.5	20.0

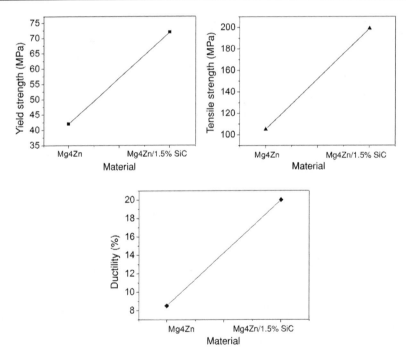

5.5.3.3. Mg6Zn Reinforced with 50 nm SiC (by Melt Casting—Ultrasonic Cavitation)

Processing method	Melt casting (ultrasonic cavitation processing)
Matrix materials used	Mg6Zn
Size of SiC used	Average 50 nm

(a) As-Cast Condition
Refer to Table 5.38.

TABLE 5.38. Characteristics of Mg6Zn and Mg6Zn/SiC composites [46].

Material	Mg6Zn	Mg6Zn/1.5% SiC
Amount of SiC (wt%)	—	1.5
Yield strength (MPa)	54	76
Tensile strength (MPa)	150	201
Elongation (%)	7.7	8

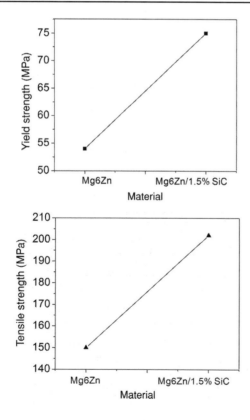

(b) Heat-Treated Condition (T5)

- Heat treated at 330°C for 2 h, followed by air cooling, then at 180°C for 16 h, followed by air cooling.

Refer to Table 5.39.

TABLE 5.39. Characteristics of Mg6Zn and Mg6Zn/SiC composites [46].

Material	Mg6Zn	Mg6Zn/1.5% SiC
Amount of SiC (wt%)	—	1.5
Yield strength (MPa)	125	168
Tensile strength (MPa)	198	235
Elongation (%)	3.5	3.0

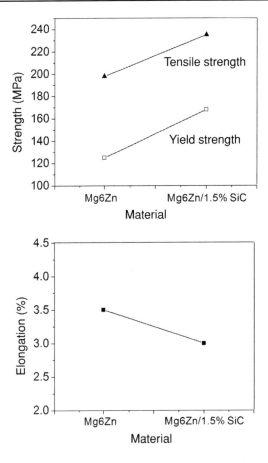

5.5.3.4. Mg Reinforced with 45–55 nm SiC (by Powder Metallurgy)

Processing method	Powder metallurgy route
Matrix material used	Pure Mg (98.5% purity)
Size of Mg used	60–300 μm
Reinforcement used	β-SiC (97.5+% purity)
Size of SiC used	45–55 nm
Extrusion temperature	350°C
Extrusion ratio	25:1

(a) Without Sintering Process
Refer to Table 5.40.

TABLE 5.40. Characteristics of Mg and Mg/SiC [47].

Material	Mg-US	Mg/0.35 SiC-US	Mg/0.5 SiC-US	Mg/1.0 SiC-US
Wt% of SiC (vol.%)	—	0.65 (0.35)	0.92 (0.50)	1.84 (1.00)
Density (g/cm^3)	1.725	1.721	1.723	1.719
Porosity (%)	0.86	1.39	1.41	2.02
CTE ($\times 10^{-6}$ /°C)	30.1	29.3	28.6	28.5
0.2% YS (MPa)	106 ± 7	116 ± 12	107 ± 10	125 ± 2
UTS (MPa)	160 ± 8	169 ± 17	161 ± 11	181 ± 4
Ductility (%)	5.8 ± 1.0	5.2 ± 1.4	6.5 ± 0.2	6.1 ± 0.9
Microhardness (HV)—matrix	35 ± 1	38 ± 1	40 ± 1	41 ± 2

US, unsintered.

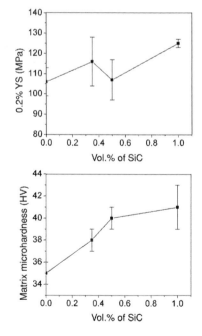
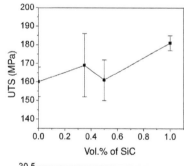
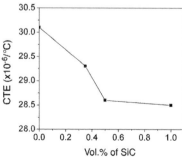

(b) Hybrid Microwave-Assisted Sintering

- Sintered for 25 min

Refer to Table 5.41.

TABLE 5.41. Characteristics of Mg and Mg/SiC [47].

Material	Mg-MW	Mg/0.35 SiC-MW	Mg/0.5 SiC-MW	Mg/1.0 SiC-MW
Wt% of SiC (vol.%)	—	0.65 (0.35)	0.92 (0.50)	1.84 (1.00)
Density (g/cm^3)	1.731	1.735	1.739	1.753
Porosity (%)	0.52	0.58	0.48	0.11
CTE ($\times 10^{-6}$ /°C)	29.1	28.3	28.3	28.1
0.2% YS (MPa)	125 ± 15	132 ± 14	144 ± 12	157 ± 22
UTS (MPa)	172 ± 12	194 ± 11	194 ± 10	203 ± 22
Ductility (%)	5.8 ± 0.9	6.3 ± 1.0	7.0 ± 2.0	7.6 ± 1.5
Microhardness (HV)—matrix	39 ± 2	40 ± 1	42 ± 1	43 ± 2

MW, hybrid microwave-assisted sintering.

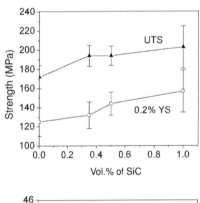

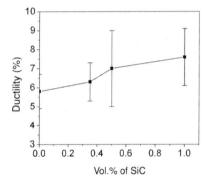

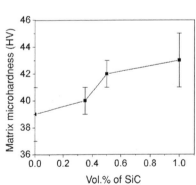

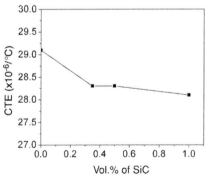

5.5.4. Addition of Hybrid Reinforcement (with SiC)

5.5.4.1. Mg Reinforced with SiC of Different Sizes (by Powder Metallurgy—Microwave Sintering)

Processing method   Powder metallurgy route
Sintering method   Hybrid microwave-assisted sintering
Matrix material used   Pure Mg (98.5% purity)
Size of Mg used   60–300 μm
Size of SiC used   25 μm and 50 nm
Sintering time   25 min
Extrusion temperature   350°C
Extrusion ratio   25:1

Refer to Table 5.42.

TABLE 5.42. Characteristics of Mg and Mg/SiC [48].

Material	(1) Mg	(2) Mg 1% nm SiC	(3) Mg 10% μm SiC	(4) Mg Hybrid
Vol.% of SiC	—	1% (50 nm)	10% (25 μm)	9% (25 μm) 1% (50 nm)
Density (g/cm³)	1.731	1.753	1.865	1.855
Porosity (%)	0.52	0.11	1.22	1.75
CTE ($\times 10^{-6}$/°C)	29.1 ± 1.2	28.1 ± 0.9	25.6 ± 1.9	24.9 ± 0.5
0.2% YS (MPa)	125 ± 15	157 ± 22	140 ± 2	156 ± 7
UTS (MPa)	172 ± 12	203 ± 22	165 ± 2	185 ± 11
Ductility (%)	5.8 ± 0.9	7.6 ± 1.5	1.5 ± 0.8	0.6 ± 0.1
Microhardness (HV) – matrix	38.6 ± 1.5	43.2 ± 2.0	44.3 ± 0.5	50.9 ± 1.0

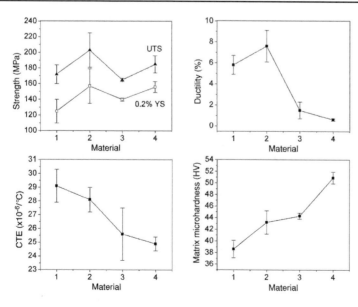

5.5.4.2. Mg Reinforced with 50 nm Al₂O₃ and 50 nm SiC (by Powder Metallurgy—Microwave Sintering)

Processing method   Powder metallurgy route
Sintering method   Hybrid microwave-assisted sintering
Matrix material used   Pure Mg (98.5% purity)
Size of Mg used   60–300 μm
Size of SiC and Al₂O₃ used   50 nm
Sintering time   27 min
Extrusion temperature   350°C
Extrusion ratio   25:1

Refer to Table 5.43.

TABLE 5.43. Characteristics of Mg and Mg-based composites [49].

Material	(1) Mg	(2) Mg/1 SiC	(3) Mg/0.5 SiC 0.5 Al₂O₃	(4) Mg/0.3 SiC 0.7 Al₂O₃
Vol.% of reinforcement (type)	—	1.0% (SiC)	0.5% (SiC) 0.5% (Al₂O₃)	0.3% (SiC) 0.7% (Al₂O₃)
Density (g/cm³)	1.7345	1.7414	1.7410	1.7407
Porosity (%)	0.32	0.38	0.44	0.47
CTE ($\times 10^{-6}$/°C)	29.3 ± 0.7	28.0 ± 0.7	27.6 ± 0.7	26.7 ± 0.5
0.2% YS (MPa)	119 ± 8	131 ± 12	156 ± 7	165 ± 1
UTS (MPa)	169 ± 4	182 ± 9	197 ± 2	206 ± 5
Failure strain (%)	5.5 ± 1.6	5.0 ± 0.5	4.6 ± 2.1	4.2 ± 1.8
Macrohardness (HR15T)	43.3 ± 1.7	45.8 ± 1.2	47.0 ± 1.0	47.9 ± 0.6
Microhardness (HV)—matrix	39.1 ± 0.8	43.1 ± 0.7	46.1 ± 0.9	48.0 ± 0.5

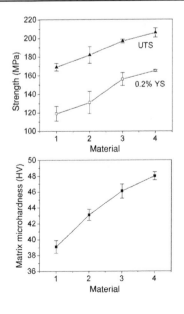
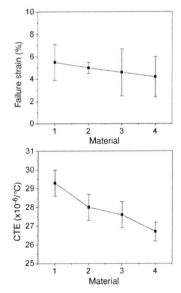

5.5.4.3. Mg Reinforced with 50 nm SiC and MWCNT (by Powder Metallurgy—Microwave Sintering)

Processing method	Powder metallurgy route
Sintering method	Hybrid microwave-assisted sintering
Matrix material used	Pure Mg (98.5% purity)
Size of Mg used	60–300 μm
Size of SiC used	50 nm
MWCNT used	40–70 nm (outer diameter); >95% purity
Sintering time	25 min
Extrusion temperature	350°C
Extrusion ratio	25:1

Refer to Table 5.44.

TABLE 5.44. Characteristics of Mg and Mg-based composites [50].

	(1)	(2)	(3)	(4)	(5)
		Mg/0.3 CNT	Mg/0.5 CNT	Mg/0.7 CNT	
Material	Mg	0.7 SiC	0.5 SiC	0.3 SiC	Mg/1 CNT
Wt% of reinforcement (type)	—	0.3% (SiC) 0.7% (MWCNT)	0.5% (SiC) 0.5% (MWCNT)	0.7% (SiC) 0.3% (MWCNT)	1.0% (MWCNT)
Density (g/cm^3)	1.737	1.742	1.740	1.739	1.736
Porosity (%)	0.17	0.23	0.24	0.23	0.29
CTE ($\times 10^{-6}$/°C)	29.0 ± 1.1	28.0 ± 0.4	28.1 ± 0.5	28.3 ± 0.6	28.7 ± 0.3
0.2% YS (MPa)	112 ± 8	153 ± 4	152 ± 1	140 ± 7	117 ± 6
UTS (MPa)	156 ± 2	195 ± 5	189 ± 3	183 ± 8	154 ± 3
Failure strain (%)	5.9 ± 1.2	3.3 ± 0.7	2.3 ± 0.6	2.1 ± 0.5	1.5 ± 0.3
Microhardness (HV)—matrix	41 ± 1	46 ± 1	45 ± 1	44 ± 1	43 ± 1

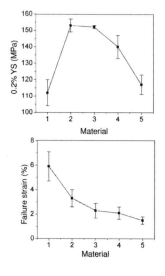
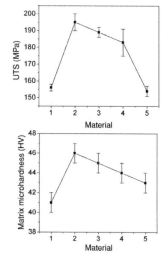

5.6. MAGNESIUM-BASED COMPOSITES WITH Y_2O_3

Limited studies have been carried out so far to investigate the integration effects of Y_2O_3 with magnesium. Disintegrated melt deposition and powder metallurgy methods have been employed. Y_2O_3 is used at nanolength scale as singular and hybrid reinforcement (with Cu and Ni).

For the powder metallurgy processed samples, the end properties depends on

(i) the sintering type,
(ii) the heating rate during microwave sintering, and
(iii) the extrusion ratio employed.

As Y_2O_3 has the capability to positively influence the properties of magnesium, it is expected that researchers will pick up this reinforcement even more constructively in the time to come.

5.6.1. Addition of Nanosize Y_2O_3

5.6.1.1. Mg Reinforced with 29 nm Y_2O_3 (by Disintegrated Melt Deposition)

Processing method	Disintegrated melt deposition
Matrix material used	Pure Mg (purity 99.9%)
Size of Y_2O_3 used	29 nm
Extrusion temperature	250°C
Extrusion ratio	20.25:1

Refer to Table 5.45.

TABLE 5.45. Characteristics of Mg and Mg/Y$_2$O$_3$ [55].

Material	Mg	Mg/0.22 Y$_2$O$_3$	Mg/0.66 Y$_2$O$_3$	Mg/1.11 Y$_2$O$_3$
Vol.% of Y$_2$O$_3$ (wt%)	0.00	0.22 (0.6)	0.66 (1.9)	1.11 (3.1)
Density (g/cm^3)	1.7397	1.7472	1.7598	1.7730
CTE ($\times 10^{-6}$/°C)	0.02	0.00	0.10	0.19
Macrohardness (HR15T)	37 ± 1	56 ± 1	58 ± 0	49 ± 0
Microhardness (HV)	40 ± 0	51 ± 0	56 ± 0	52 ± 1
0.2% YS (MPa)	97 ± 2	218 ± 2	312 ± 4	—
UTS (MPa)	173 ± 1	277 ± 5	318 ± 2	205 ± 3
Ductility (%)	7.4 ± 0.2	12.7 ± 1.3	6.9 ± 1.6	1.7 ± 0.5
WoF (J/m^3)	11.1 ± 0.3	29.6 ± 3.5	18.2 ± 4.7	1.9 ± 0.7

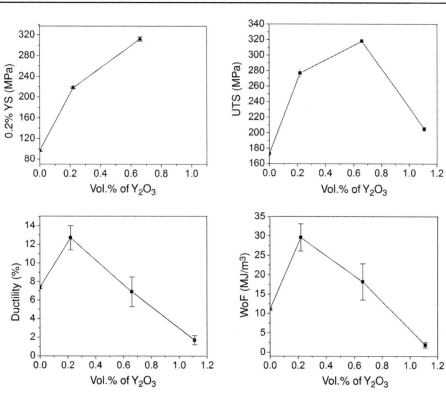

5.6.1.2. Mg Reinforced with 29 nm Y₂O₃ (by Powder Metallurgy—Conventional Sintering)

Processing method   Powder metallurgy route
Sintering method   Conventional tube furnace sintering
Matrix material used   Pure Mg ($\geq$98.5% purity)
Size of Mg used   60–300 μm
Size of Y_2O_3 used   29 nm
Extrusion temperature   250°C
Extrusion ratio   20.25:1

Refer to Table 5.46.

TABLE 5.46. Characteristics of Mg and Mg/Y₂O₃ [56].

Material	Mg	Mg/0.22Y₂O₃	Mg/0.66Y₂O₃	Mg/1.11Y₂O₃
Wt% of Y₂O₃ (vol.%)	—	0.6 (0.22)	1.9 (0.66)	3.1 (1.11)
Density (g/cm³)	1.7387	1.7455	1.7555	1.7675
Porosity (%)	0.08	0.1	0.35	0.49
0.2% YS (MPa)	132 ± 7	156 ± 1	151 ± 2	153 ± 3
UTS (MPa)	193 ± 2	211 ± 1	202 ± 2	195 ± 2
Ductility (%)	4.2 ± 0.1	15.8 ± 0.7	12.0 ± 1.0	9.1 ± 0.2
Macrohardness (HR15T)	43 ± 0	44 ± 1	46 ± 1	49 ± 0
Microhardness (HV)—matrix	37 ± 0	38 ± 0	38 ± 1	51 ± 1

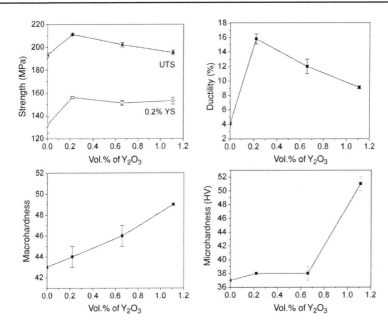

5.6.1.3. Mg Reinforced with 32–36 nm Y_2O_3 (by Disintegrated Melt Deposition)

Processing method	Disintegrated melt deposition
Matrix material used	Pure Mg (99.9% purity)
Size of Y_2O_3 used	32–36 nm
Extrusion temperature	350°C
Extrusion ratio	20.25:1

Refer to Table 5.47.

TABLE 5.47. Characteristics of Mg and Mg/Y_2O_3 [57].

Material	Mg	Mg/0.5 Y_2O_3	Mg/1.0 Y_2O_3	Mg/2.0 Y_2O_3
Vol.% of Y_2O_3	—	0.5	1.0	2.0
Density (g/cm³)	1.738	1.742	1.743	1.749
CTE ($\times 10^{-6}$/°C)	28.7 ± 0.6	27.8 ± 0.4	27.0 ± 0.8	26.8 ± 0.5
0.2% YS (MPa)	126 ± 7	141 ± 7	151 ± 5	162 ± 10
UTS (MPa)	192 ± 5	223 ± 5	222 ± 4	227 ± 11
Ductility (%)	8.0 ± 1.6	8.5 ± 1.6	6.8 ± 0.5	7.0 ± 0.5

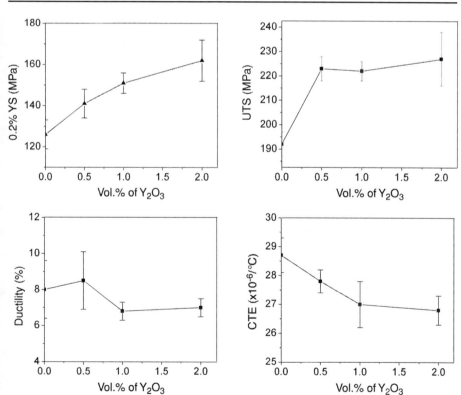

5.6.1.4. Mg Reinforced with 30–50 nm Y₂O₃ (by Powder Metallurgy—Microwave Sintering)

Processing method	Powder metallurgy route
Sintering method	Hybrid microwave-assisted sintering
Matrix material used	Pure Mg (98.5% purity)
Size of Mg used	60–300 μm
Size of Y_2O_3 used	30–50 nm
Extrusion temperature	350°C

(a) Effect of Amount of Y_2O_3 Addition

Sintering time	13 min
Extrusion ratio	25:1

Refer to Table 5.48.

TABLE 5.48. Characteristics of Mg and Mg/Y₂O₃ [58].

Material	Mg	Mg/0.17 Y_2O_3	Mg/0.70 Y_2O_3
Wt% of Y_2O_3 (vol.%)	—	0.5 (0.17)	2.0 (0.7)
Density (g/cm³)	1.74	1.73	1.76
Porosity (%)	0.13	0.87	0.35
CTE ($\times 10^{-6}$/°C)	28.2 ± 0.0	21.3 ± 0.1	20.8 ± 0.6
0.2% YS (MPa)	134 ± 7	144 ± 2	157 ± 10
UTS (MPa)	193 ± 1	214 ± 4	244 ± 1
Ductility (%)	7.5 ± 2.5	8.0 ± 2.8	8.6 ± 1.2
Work of fracture (MJ/m³)	12.9 ± 4.8	16.6 ± 4.2	21.8 ± 3.1
Microhardness (HV)—matrix	37 ± 2	38 ± 0	45 ± 2

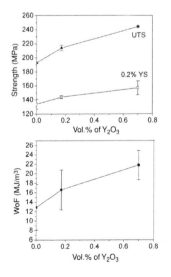
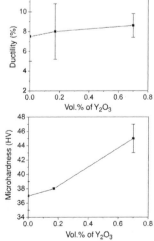

(b) Effect of Heating Rate

Heating rate during sintering   49°C/min, 20°C/min
Sintering time   13 min
Extrusion ratio   25:1

Refer to Table 5.49.

TABLE 5.49. Characteristics of Mg and Mg/Y₂O₃ [59].

	(1)	(2)	(3)	(4)
Material	Mg	Mg	Mg/0.7 Y$_2$O$_3$	Mg/0.7 Y$_2$O$_3$
Heating rate (°C/min)	49	20	49	20
Wt% of Y$_2$O$_3$ (vol.%)	—	2.0 (0.7)	—	2.0 (0.7)
Density (g/cm^3)	1.738	1.734	1.757	1.754
Porosity (%)	0.13	0.33	0.35	0.49
0.2% YS (MPa)	134 ± 7	116 ± 17	157 ± 10	111 ± 10
UTS (MPa)	193 ± 1	186 ± 21	244 ± 1	175 ± 8
Failure strain (%)	6.9 ± 2.5	11.3 ± 1.0	9.1 ± 0.6	9.2 ± 0.5

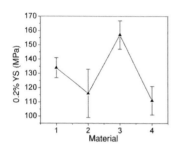
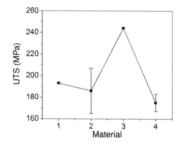
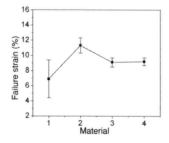

(c) Effect of Extrusion Ratio
Extrusion ratio   25:1, 19:1, 12:1
Refer to Table 5.50.

TABLE 5.50. Characteristics of Mg and Mg/Y$_2$O$_3$ [60].

Material	Mg	Mg	Mg	Mg/Y$_2$O$_3$	Mg/Y$_2$O$_3$	Mg/Y$_2$O$_3$
Extrusion ratio	12:1	19:1	25:1	12:1	19:1	25:1
Wt% of Y$_2$O$_3$	—	—	—	2.0	2.0	2.0
0.2% YS (MPa)	88 ± 8	99 ± 2	134 ± 7	69 ± 5	74 ± 3	157 ± 10
UTS (MPa)	147 ± 7	165 ± 4	193 ± 1	119 ± 6	128 ± 8	244 ± 1
Failure strain (%)	8.0 ± 1.3	10.3 ± 4.7	6.9 ± 2.6	6.7 ± 0.7	9.3 ± 0.5	9.1 ± 0.6
Work of fracture (MJ/m^3)	11.2 ± 1.9	16.3 ± 1.5	12.9 ± 4.8	7.4 ± 0.7	11.2 ± 1.1	21.8 ± 3.1
Microhardness (HV)—matrix	33 ± 1	34 ± 1	37 ± 2	35 ± 2	36 ± 2	45 ± 2

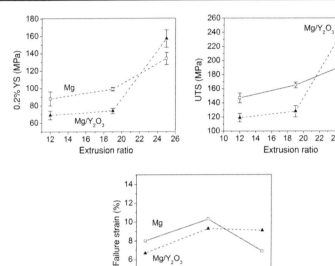

5.6.2. Addition of Hybrid Reinforcements (with Y$_2$O$_3$)

5.6.2.1. Mg Reinforced with Y$_2$O$_3$ and Nanosize Cu (by Powder Metallurgy—Microwave Sintering)

Processing method	Powder metallurgy route
Sintering method	Hybrid microwave-assisted sintering
Matrix material used	Pure Mg (98.5% purity)
Size of Mg used	60–300 μm
Size of Y$_2$O$_3$ used	30–50 nm
Size of Cu used	25 nm
Extrusion temperature	350°C
Sintering time	13 min
Extrusion ratio	25:1

Refer to Table 5.51.

TABLE 5.51. Characteristics of Mg and Mg/Y$_2$O$_3$+Cu composites [61].

Material	(1) Mg	(2) Mg/(0.7 Y$_2$O$_3$ + 0.3 Cu)	(3) Mg/(0.7 Y$_2$O$_3$ + 0.6 Cu)
Vol.% of Y$_2$O$_3$ (Vol.% of Cu)	—	0.7 (0.3)	0.7 (0.6)
Density (g/cm^3)	1.74	1.78	1.79
Porosity (%)	0.13	0.45	0.77
0.2% YS (MPa)	134 ± 7	215 ± 20	179 ± 7
UTS (MPa)	193 ± 1	270 ± 22	231 ± 13
Failure strain (%)	6.9 ± 2.5	11.1 ± 1.0	11.1 ± 0.7
Work of fracture (MJ/m^3)	12.9 ± 4.8	29.8 ± 2.7	25.4 ± 0.9

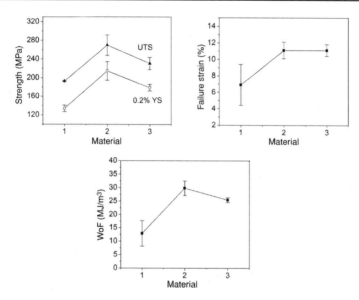

5.6.2.2. Mg Reinforced with Y$_2$O$_3$ and Nanosize Ni (by Powder Metallurgy—Microwave Sintering)

Processing method	Powder metallurgy route
Sintering method	Hybrid Microwave-assisted sintering
Matrix material used	Pure Mg (98.5% purity)
Size of Mg used	60–300 μm
Size of Y$_2$O$_3$ used	30–50 nm
Size of Ni used	20 nm
Extrusion temperature	350°C
Sintering time	13 min
Extrusion ratio	25:1

Refer to Table 5.52.

TABLE 5.52. Characteristics of Mg and Mg/Y$_2$O$_3$+Ni composites [62].

Material	(1) Mg	(2) Mg/ 0.7 Y$_2$O$_3$	(3) Mg/ (0.7 Y$_2$O$_3$ + 0.3 Ni)	(4) Mg/ (0.7 Y$_2$O$_3$ + 0.6 Ni)	(5) Mg/ (0.7 Y$_2$O$_3$ + 1.0 Ni)
Vol.% of Y$_2$O$_3$ (vol.% of Ni)	—	0.7 (0.0)	0.7 (0.3)	0.7 (0.6)	0.7 (1.0)
Density (g/cm^3)	1.74	1.76	1.78	1.80	1.83
Porosity (%)	0.13	0.35	0.34	0.21	0.30
Microhardness (HV)	37 ± 2	45 ± 2	54 ± 4	60 ± 4	63 ± 4
0.2% YS (MPa)	134 ± 7	157 ± 10	221 ± 7	232 ± 8	228 ± 8
UTS (MPa)	193 ± 1	244 ± 1	262 ± 6	272 ± 2	271 ± 6
Failure strain (%)	6.9 ± 2.5	9.1 ± 0.6	9.0 ± 0.9	9.5 ± 0.9	5.5 ± 0.7
Work of fracture (MJ/m^3)	12.9 ± 4.8	21.8 ± 3.1	23.7 ± 2.1	25.9 ± 2.3	15.4 ± 2.3

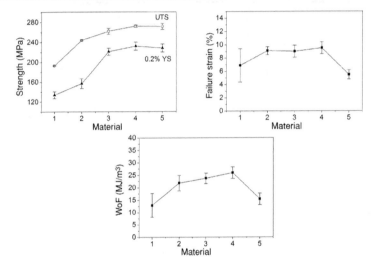

5.7. MAGNESIUM-BASED COMPOSITES WITH ZrO$_2$

Only two studies have been reported on Mg/ZrO$_2$ composites. Processing methods include disintegrated melt deposition (solidification-based) and powder metallurgy. Length scale of ZrO$_2$ particulates is in the nanometric range (<100 nm). For the similar amounts, it was found that solidification method was more effective in increasing strength, while powder metallurgy method was more effective in increasing ductility.

5.7.1. Addition of Nanosize ZrO$_2$

5.7.1.1. Mg Reinforced with 29–68 nm ZrO$_2$ (by Disintegrated Melt Deposition)

Processing method	Disintegrated melt deposition
Matrix material used	Pure Mg (99.9% purity)
Size of ZrO$_2$ used	29–68 nm (99.5% purity)
Extrusion temperature	250°C
Extrusion ratio	20.25:1

Refer to Table 5.53.

TABLE 5.53. Characteristics of Mg and Mg/ZrO$_2$ [63].

Material	Mg	Mg/0.22 ZrO$_2$	Mg/0.66 ZrO$_2$	Mg/1.11 ZrO$_2$
Wt% of ZrO$_2$ (vol.%)	—	0.74 (0.22)	2.20 (0.66)	3.66 (1.11)
Density (g/cm^3)	1.7397	1.7466	1.7644	1.7855
Porosity (%)	0.02	0.15	0.17	0.03
0.2% YS (MPa)	97 ± 2	186 ± 2	221 ± 5	216 ± 4
UTS (MPa)	173 ± 1	248 ± 4	271 ± 6	250 ± 6
Ductility (%)	7.4 ± 0.2	4.7 ± 0.2	4.8 ± 0.7	3.0 ± 0.2
WoF (J/m^3)	11.1 ± 0.3	9.8 ± 0.9	10.8 ± 1.5	6.1 ± 0.6
Macrohardness (HR15T)	37.1 ± 0.7	58.8 ± 0.6	61.9 ± 0.7	62 8 ± 0.5
Microhardness (HV)—matrix	40.0 ± 0.2	47.1 ± 0.6	51.0 ± 0.6	54.7 ± 0.7

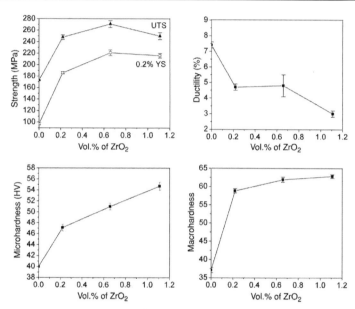

5.7.1.2. Mg Reinforced with 29–68 nm ZrO$_2$ (by Powder Metallurgy—Conventional Sintering)

Processing method	Powder metallurgy route
Sintering method	Conventional tube furnace sintering
Matrix material used	Pure Mg (98.5% purity)
Size of Mg used	60–300 μm
Size of ZrO$_2$ used	29–68 nm (99.5% purity)
Extrusion temperature	250°C
Extrusion ratio	20.25 : 1

Refer to Table 5.54.

TABLE 5.54. Characteristics of Mg and Mg/ZrO$_2$ [64].

Material	Mg	Mg/0.22 ZrO$_2$	Mg/0.66 ZrO$_2$	Mg/1.11 ZrO$_2$
Wt% of ZrO$_2$ (vol.%)	—	0.7 (0.22)	2.2 (0.66)	3.7 (1.11)
Density (g/cm^3)	1.7387	1.7436	1.7599	1.7829
Porosity (%)	0.08	0.32	0.42	0.18
0.2% YS (MPa)	132 ± 7	140 ± 3	163 ± 3	146 ± 1
UTS (MPa)	193 ± 2	160 ± 8	202 ± 6	199 ± 5
Ductility (%)	4.2 ± 0.1	6.4 ± 1.5	11.4 ± 0.9	10.8 ± 1.3
WoF (J/m^3)	7.1 ± 0.3	12.2 ± 2.8	27.2 ± 1.9	27.2 ± 4.6
Microhardness (HV)—matrix	37 ± 0	38 ± 0	44 ± 1	46 ± 1

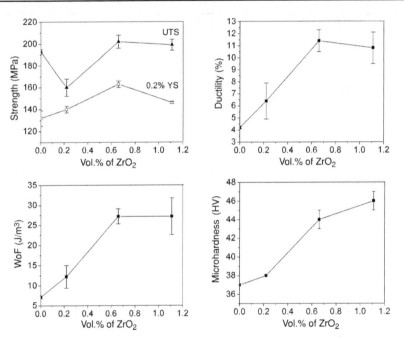

5.8. MAGNESIUM-BASED COMPOSITES WITH CNT

CNTs are used to a limited extent as reinforcement in magnesium and its alloys. Magnesium matrices include pure magnesium, AZ31, and AZ91D. The processing methods employed were powder metallurgy, disintegrated melt deposition, and melt stirring. Using both disintegrated melt deposition and powder metallurgy methods, the addition of CNTs to pure Mg led to an increase in ductility, while the increase in 0.2% yield strength was insignificant.

For the case of AZ31 alloy, addition of 1 vol.% CNTs led to a simultaneous improvement in strength and ductility under both tensile and compressive loadings. For the case of AZ91D, the overall compressive behavior increased with the addition of CNTs.

5.8.1. Addition of MWCNTs

5.8.1.1. Mg Reinforced with MWCNTs (by Disintegrated Melt Deposition)

Processing method	Disintegrated melt deposition
Matrix material used	Pure Mg (>99.9% purity)
Diameter of MWCNT used	20 nm (average)
Extrusion temperature	350°C
Extrusion ratio	20.25:1

Refer to Table 5.55.

TABLE 5.55. Characteristics of Mg and Mg/CNT [65, 66].

Material	Mg	Mg/0.3 CNT	Mg/1.3 CNT	Mg/1.6 CNT	Mg/2.0 CNT
Wt% of MWCNT	—	0.3	1.3	1.6	2.0
Density (g/cm^3)	1.738	1.731	1.730	1.731	1.728
0.2% YS (MPa)	126 ± 7	128 ± 6	140 ± 2	121 ± 5	122 ± 7
UTS (MPa)	192 ± 5	194 ± 9	210 ± 4	200 ± 3	198 ± 8
Ductility (%)	8.0 ± 1.6	12.7 ± 2.0	13.5 ± 2.7	12.2 ± 1.7	7.7 ± 1.0
Macrohardness (HR15T)	45 ± 1	48 ± 1	46 ± 1	42 ± 1	39 ± 1

MWCNT, multiwalled carbon nanotubes.

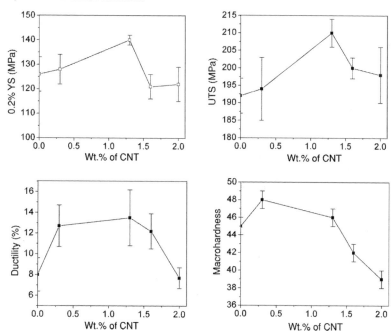

5.8.1.2. Mg Reinforced with MWCNTs (by Powder Metallurgy—Conventional Sintering)

Processing method	Powder metallurgy route
Sintering method	Conventional tube furnace sintering
Matrix material used	Pure Mg (98.5% purity)
Diameter of MWCNT used	20 nm (average)
Extrusion temperature	350°C
Extrusion ratio	20.25:1

Refer to Table 5.56.

TABLE 5.56. Characteristics of Mg and Mg/CNT [65, 67].

Material	Mg	Mg/0.06 CNT	Mg/0.18 CNT	Mg/0.30 CNT
Wt% of MWCNT	—	0.06	0.18	0.30
Density (g/cm^3)	1.738	1.738	1.737	1.736
Porosity (%)	0.12	0.25	0.83	1.15
CTE ($\times 10^{-6}$/°C)	28.6	27.2	26.2	25.9
0.2% YS (MPa)	127 ± 5	133 ± 2	138 ± 4	146 ± 5
UTS (MPa)	205 ± 4	203 ± 1	206 ± 7	210 ± 6
Ductility (%)	9 ± 2	12 ± 1	11 ± 1	8 ± 1
WoF (MJ/m^3)	18.9 ± 4.2	22.4 ± 1.5	20.8 ± 2.1	15.8 ± 1.8
Macrohardness (HR15T)	45 ± 0	44 ± 0	44 ± 1	44 ± 0

MWCNT, multiwalled carbon nanotubes.

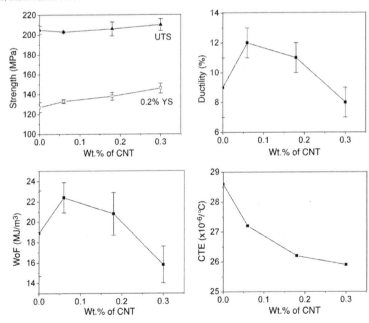

5.8.1.3. AZ91D Reinforced with MWCNTs (by Powder Metallurgy— Mechanical Milling)

Processing method	Powder metallurgy route—mechanical milling
Matrix material used	AZ91D
Size of AZ91D	$\leq 100\ \mu m$
Length of MWCNT used	$\sim 5\ \mu m$
Extrusion temperature	450°C
Extrusion ratio	9:1
Solution treatment	410°C for 15 min
Artificial aging	200°C for 15 h

Refer to Table 5.57.

TABLE 5.57. Characteristics of AZ91D and AZ91D/CNT [68].

Material	AZ91D	AZ91D/0.5% CNT	AZ91D/1% CNT	AZ91D/3% CNT	AZ91D/5% CNT
% of MWCNT	—	0.5	1.0	3.0	5.0
Density (g/cm³)	1.80	1.82	1.83	1.84	1.86
Elastic modulus (GPa)	40 ± 2	43 ± 3	49 ± 3	51 ± 3	51 ± 4
0.2% proof stress (MPa)	232 ± 6	281 ± 6	295 ± 5	284 ± 6	277 ± 4
Tensile stress (MPa)	315 ± 5	383 ± 7	388 ± 11	361 ± 9	307 ± 10
Ductility (%)	14 ± 3	6 ± 2	5 ± 2	3 ± 2	1 ± 0.5

MWCNT, multiwalled carbon nanotubes.

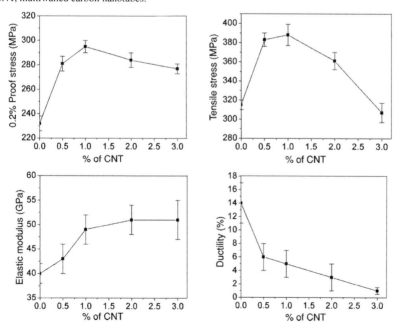

5.8.1.4. AZ91D Reinforced with MWCNTs (by Melt Stirring Method)

Processing method	Melt stirring method
Matrix material used	AZ91D (chips)
Diameter of MWCNT used	5–20 nm
Predispersion of MWCNTs	Block copolymer Disperbyk-2150 dissolved in ethanol solution
	Ultrasonic batch (15 min, room temperature)
	Homogeneous stirring (30 min, 250 rpm)
Predispersion of Mg alloy chips in MWCNT solution	Evaporation of ethanol
	Homogeneous stirring (250 rpm)
	Evaporation of ethanol

Refer to Table 5.58.

TABLE 5.58. Characteristics of AZ91D and AZ91D/0.1 wt% CNT [69].

Material	AZ91D	AZ91D + 0.1 wt% MWCNTs
2% Compressive yield strength (MPa)	248	272
Ultimate compressive strength (MPa)	344	412
Compression at failure (%)	18	24.4

MWCNT, multiwalled carbon nanotubes.

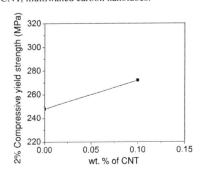

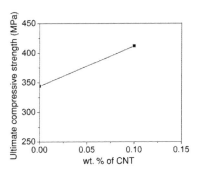

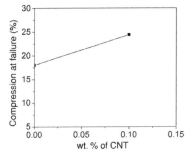

5.8.1.5. AZ31 Reinforced with CNTs (by Disintegrated Melt Deposition)

Processing method	Disintegrated melt deposition (DMD)
Matrix material used	AZ31 rod (nominally 2.50–3.50 wt% Al, 0.60–1.40 wt% Zn, 0.15–0.40 wt% Mn, 0.10 wt% Si, 0.05 wt% Cu, 0.01 wt% Fe, 0.01 wt% Ni, balance Mg)
CNT used	94.7% (purity), 40–70 nm (outer diameter)
Extrusion temperature	350°C
Extrusion ratio	20.25:1

(a) Tensile Test
Refer to Table 5.59.

TABLE 5.59. Tensile test results of AZ31 and AZ31/1.0 vol.% CNT [70].

Material	AZ31	AZ31 + 1.0 vol.% CNT
0.2% Yield Strength (MPa)	172 ± 15	190 ± 13
Ultimate Tensile Strength (MPa)	263 ± 12	307 ± 10
Failure Strain (%)	10.4 ± 3.9	17.5 ± 2.6
WoF (MJ/m^3)	26 ± 9	50 ± 8
Microhardness (HV)—matrix	64 ± 4	95 ± 4

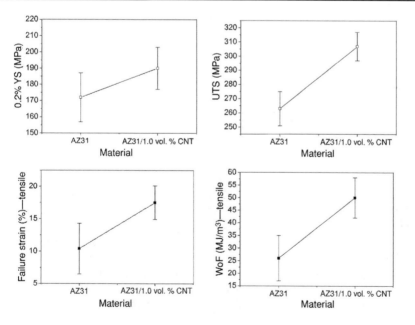

(b) Compression Test
Refer to Table 5.60.

TABLE 5.60. Compression test results of AZ31 and AZ31/1.0vol.%CNT [70].

Material	AZ31	AZ31 + 1.0 vol.% CNT
0.2% CYS (MPa)	93 ± 9	147 ± 13
UCS (MPa)	486 ± 4	501 ± 25
Failure strain (%)—compressive	19.7 ± 7.2	20.6 ± 2.8
WOF (MJ/m^3)—compressive	76 ± 14	89 ± 16

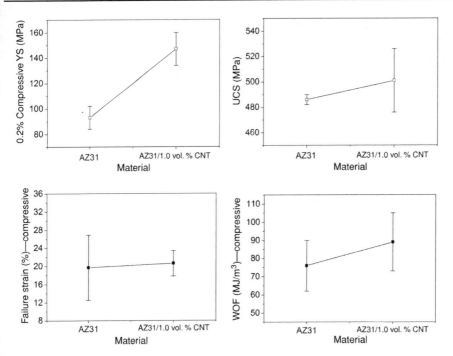

5.9. MAGNESIUM-BASED COMPOSITES WITH METALLIC ADDITIONS

A number of attempts have been made in the recent years to integrate high melting point metals with magnesium. Processing methods include both solidification and powder metallurgy methods. Examples of metallic reinforcements include copper, nickel, titanium, molybdenum, aluminum, and iron. Results indicate that strength (Cu and Ni), ductility (Ti and Mo), and modulus (interconnected reinforcement) can be enhanced depending on the choice of reinforcement.

Much more efforts are required to investigate the mechanisms of action of metallic reinforcements in changing the properties of magnesium and its alloys.

5.9.1. Addition of Micrometer-Size Copper

5.9.1.1. Mg Reinforced with 8–11 μm Cu (by Disintegrated Melt Deposition)

Processing method	Disintegrated melt deposition
Matrix material used	Pure Mg (> 99.9% purity)
Size of Cu used	8–11 μm
Extrusion temperature	350°C
Extrusion ratio	20.25:1

Refer to Table 5.61.

TABLE 5.61. Characteristics of Mg and Mg/Cu [72, 73].

Material	Mg	Mg/10.1 Cu	Mg/18.0 Cu	Mg/26.6 Cu
Wt% of Cu (vol.%)	—	10.1 (2.1)	18.0 (4.1)	26.6 (6.6)
Density (g/cm³)	1.739	1.893	2.094	2.224
Porosity (%)	0.05	0.29	0.01	0.19
CTE ($\times 10^{-6}$/°C)	28.6 ± 0.1	27.9 ± 0.8	28.1 ± 0.1	25.7 ± 1.0
Elastic modulus (GPa)	43 ± 1	47 ± 1	47 ± 2	53 ± 1
0.2% YS (MPa)	100 ± 4	281 ± 13	355 ± 11	—
UTS (MPa)	258 ± 16	335 ± 15	386 ± 3	433 ± 27
Ductility (%)	7.7 ± 1.2	2.5 ± 0.2	1.5 ± 0.3	1.0 ± 0.1
Macrohardness (HR15T)	57 ± 2	65 ± 1	73 ± 1	82 ± 0
Microhardness (HV)—interface	—	71 ± 5	85 ± 3	117 ± 5
Microhardness (HV)—matrix	43 ± 0	58 ± 2	66 ± 1	102 ± 2

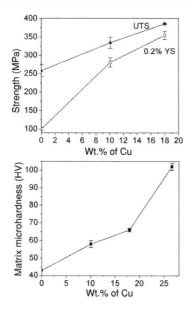

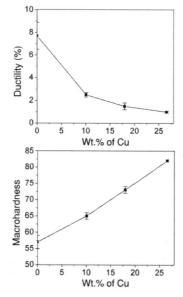

5.9.1.2. AZ91A Reinforced with 8–11 μm Cu (by Disintegrated Melt Deposition)

Processing method	Disintegrated melt deposition
Matrix material used	AZ91A
Size of Cu used	8–11 μm (99% purity)
Extrusion temperature	350°C
Extrusion ratio	20.25:1
T6 heat treatment	Solutionized at 413°C followed by aging at 168°C

Refer to Table 5.62.

TABLE 5.62. Characteristics of AZ91A and AZ91A/Cu [74].

Material	AZ91A	AZ91A/Cu
Wt% of Cu (vol.%)	—	15.54 (3.59)
Density (g/cm^3)	1.82	2.11
CTE ($\times10^{-6}$/°C)	30.7 ± 0.5	27.9 ± 0.6
Dynamic elastic modulus (GPa)	44	52
Elastic modulus^a (GPa)	43 ± 3	54 ± 1
0.2% YS (MPa)	263 ± 12	299 ± 5
UTS (MPa)	358 ± 5	382 ± 6
Failure strain (%)	7 ± 4	6 ± 1

^{a}Elastic modulus determined from tensile test.

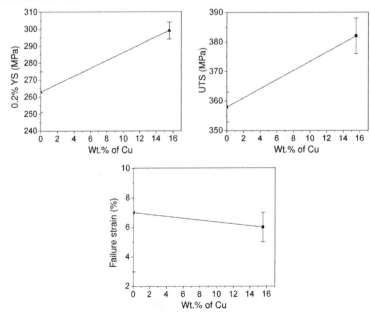

5.9.1.3. AZ91A and Mg Reinforced with 8–11 μm Cu (by Disintegrated Melt Deposition)

Processing method   Disintegrated Melt Deposition (DMD)
Matrix material used   AZ91A and pure Mg (99.9% purity)
Size of Cu used   8–11 μm (99% purity)
Extrusion temperature   350°C
Extrusion ratio   20.25:1
AZ91 samples   T6 heat treatment by solutionizing at 413°C for 1 h
                           with subsequent ageing at 168°C for 9 h.

Refer to Table 5.63.

TABLE 5.63. Characteristics of AZ91A and Mg/Cu [75].

	(1)	(2)
Material	AZ91A	Mg/Cu
Wt% of Cu (vol.%)	—	17.2 (3.9)
Density (g/cm^3)	1.820	2.076
CTE ($\times 10^{-6}$/°C)	30.7 ± 0.5	27.0 ± 0.4
0.2% YS (MPa)	272 ± 3	355 ± 8
UTS (MPa)	353 ± 0	358 ± 7
Failure strain (%)	3.7 ± 0.5	2.2 ± 0.9

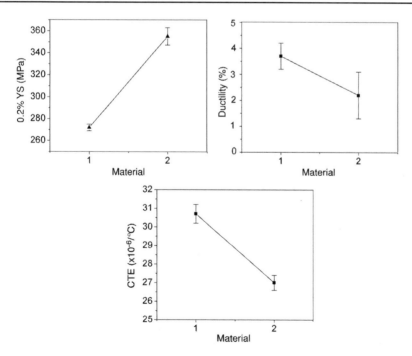

5.9.2. Addition of Nanosize Copper

5.9.2.1. Mg Reinforced with 50 nm Cu (by Powder Metallurgy—Microwave Sintering)

Processing method	Powder metallurgy route
Sintering method	Hybrid microwave-assisted sintering
Matrix material used	Pure Mg (98.5% purity)
Size of Mg used	60–300 μm
Size of Cu used	50 nm (99+% purity)
Sintering time	25 min
Extrusion temperature	350°C
Extrusion ratio	25:1

Refer to Table 5.64.

TABLE 5.64. Characteristics of Mg and Mg/Cu [76].

Material	Mg	Mg/0.3 Cu	Mg/0.6 Cu	Mg/1.0 Cu
Wt% of Cu (vol.%)	—	1.5 (0.3)	3.0 (0.6)	4.9 (1.0)
Density (g/cm^3)	1.738	1.758	1.776	1.809
Porosity (%)	0.12	0.19	0.41	0.13
CTE ($\times 10^{-6}$/°C)	28.6 ± 0.8	27.7 ± 0.3	27.3 ± 0.8	27.0 ± 0.8
Elastic modulus (GPa)	45.0 ± 0.9	57.7 ± 6.6	59.7 ± 6.7	60.1 ± 1.3
0.2% YS (MPa)	116 ± 11	188 ± 13	237 ± 24	194 ± 17
UTS (MPa)	168 ± 10	218 ± 11	286 ± 8	221 ± 17
Failure strain (%)	6.1 ± 2.0	5.9 ± 1.1	5.4 ± 1.2	2.9 ± 0.4
WoF (MJ/m^3)	11.8 ± 3.4	12.8 ± 2.0	16.5 ± 3.1	6.8 ± 1.0
Macrohardness (HR15T)	43 ± 2	56 ± 1	57 ± 1	59 ± 1
Microhardness (HV)—matrix	40 ± 1	49 ± 1	52 ± 2	60 ± 3

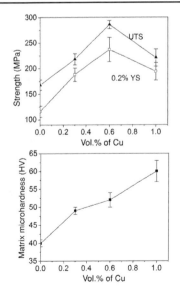

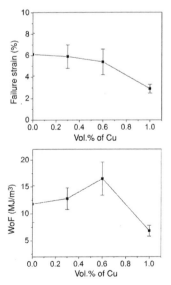

5.9.3. Addition of Nickel

5.9.3.1. Mg Reinforced with Micrometer-Size Ni (by Disintegrated Melt Deposition)

Processing method	Disintegrated melt deposition
Matrix material used	Pure Mg (>99.9% purity)
Size of Ni used	$29 \pm 19 \ \mu$m (99.9% purity)
Extrusion temperature	350°C (for Mg, Mg/7.3 Ni, and Mg/14.0 Ni)
	400°C (for Mg/24.9 Ni)
Extrusion ratio	20.25:1

Refer to Table 5.65.

TABLE 5.65. Characteristics of Mg and Mg/Ni [77].

Material	Mg	Mg/7.3 Ni	Mg/14.0 Ni	Mg/24.9 Ni
Wt% of Ni	—	7.3	14.0	24.9
Density (g/cm^3)	1.7395	1.9046	2.0677	2.3834
Porosity (%)	0.05	0.12	0.02	0.55
CTE ($\times 10^{-6}$/°C)	28.6 ± 0.1	27.5 ± 0.3	26.4 ± 0.1	20.8 ± 0.6
Elastic Modulus (GPa)	43 ± 1	47 ± 1	53 ± 1	58 ± 1
0.2% YS (MPa)	100 ± 4	337 ± 15	420 ± 27	—
UTS (MPa)	258 ± 16	370 ± 14	463 ± 4	313 ± 29^a
Ductility (%)	7.7 ± 1.2	4.8 ± 1.4	1.4 ± 0.1	0.7 ± 0.1
Macrohardness (HR15T)	57 ± 1	69 ± 1	79 ± 1	82 ± 1
Microhardness (HV—matrix	43 ± 0	65 ± 2	78 ± 1	102 ± 3
Microhardness (HV)—Mg/Ni interface	—	74 ± 3	108 ± 6	121 ± 5

^{a}Value is based on failure stress.

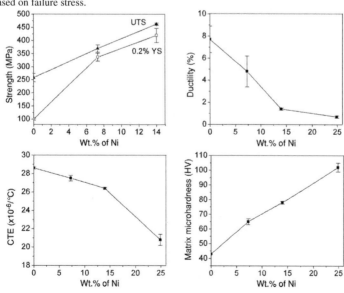

Processing method   Disintegrated melt deposition
Matrix material used   Pure Mg (>99.9% purity)
Size of Ni used   29 ± 19 μm (99.9% purity)
Extrusion temperature   350°C
Extrusion ratio   20.25:1

Refer to Table 5.66.

TABLE 5.66. Characteristics of Mg and Mg/Ni [78].

Material	Mg	Mg/14.46 Ni
Wt% of Ni	—	14.46
Density (g/cm^3)	1.739	2.049
Porosity (%)	0.05	0.12
Elastic modulus (GPa)	43 ± 1	47 ± 1
0.2% YS (MPa)	100 ± 4	370 ± 12
UTS (MPa)	258 ± 16	389 ± 5
Ductility (%)	7.7 ± 1.2	3.1 ± 0.1
Macrohardness (HR15T)	57 ± 2	76 ± 0
Microhardness (HV)—Matrix	43 ± 0	74 ± 2
Microhardness (HV)—Mg/Ni interface	—	91 ± 3

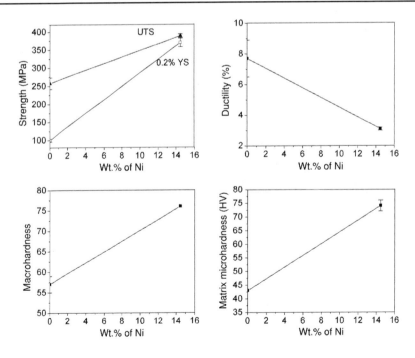

5.9.4. Addition of Titanium

5.9.4.1. Mg Reinforced with Micrometer-Size Ti (by Disintegrated Melt Deposition)

Processing method	Disintegrated melt deposition
Matrix material used	Pure Mg (>99.9% purity)
Size of Ti used	$19 \pm 10\ \mu m$ (99% purity)
Extrusion temperature	350°C
Extrusion ratio	20.25:1

Refer to Table 5.67.

TABLE 5.67. Characteristics of Mg and Mg/Ti [79].

Material	Mg	Mg/5.6Ti	Mg/9.6Ti
Wt% of Ti (vol.%)	—	5.6 (2.2)	9.6 (4.0)
Density (g/cm^3)	1.739	1.815	1.891
Porosity (%)	0.047	0.103	0.036
CTE ($\times 10^{-6}$/°C)	28.6 ± 0.1	27.9 ± 0.9	26.1 ± 1.9
0.2% YS (MPa)	100 ± 4	163 ± 12	154 ± 10
UTS (MPa)	258 ± 16	248 ± 9	239 ± 5
Ductility (%)	7.7 ± 1.2	11.1 ± 1.4	9.5 ± 0.3
WoF (J/m^3)	20.2 ± 2.6	25.7 ± 2.9	20.7 ± 1.4

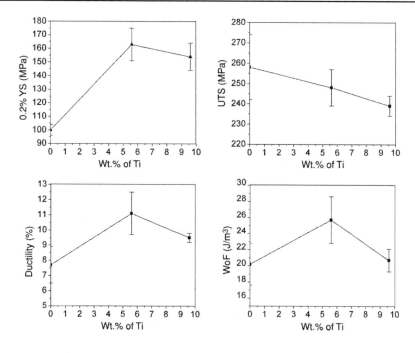

5.9.4.2. MB15 Alloy Reinforced with Micrometer-Size Ti Alloy (by Powder Metallurgy Route)

Processing method	Powder metallurgy route
Sintering method	Conventional sintering
Matrix material used	MB15 alloy (wt%: 5 Zn, 1 Zr, 0.5 Mn)
Size of MB15 alloy used	Less than 100 μm
Reinforcement material used	Ti–6Al–6V (Ti alloy)
Size of Ti alloy used	Less than 37 μm
Sintering temperature and time	530°C for 2 h
Extrusion temperature	350°C
Extrusion ratio	15:1

Refer to Table 5.68.

TABLE 5.68. Characteristics of MB15 alloy and MB15-based composites [80].

Material	MB15	MB15/Ti-6Al-6V
Vol.% of Ti-6Al-6V	—	10
Young's modulus, E (GPa)	45.5	51.6
0.2% YS (MPa)	202	278
UTS (MPa)	283	352
Ductility (%)	8.9	6.0

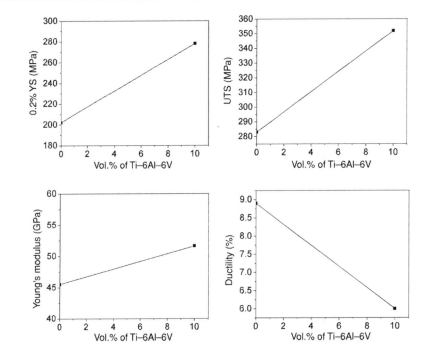

5.9.5. Addition of Molybdenum

5.9.5.1. Mg Reinforced with Micrometer-Size Mo (by Disintegrated Melt Deposition)

Processing method	Disintegrated melt deposition (DMD)
Matrix material used	Pure Mg (>99.9% purity)
Size of Mo used	325 mesh (~44 μm)
Extrusion temperature	350°C
Extrusion ratio	20.25:1

Refer to Table 5.69.

TABLE 5.69. Characteristics of Mg and Mg/Mo [82].

Material	Mg	Mg/0.7 Mo	Mg/2.0 Mo	Mg/3.6 Mo
Wt% of Mo (vol.%)	—	0.7 (0.12)	2.0 (0.34)	3.6 (0.63)
Density (g/cm^3)	1.738	1.743	1.759	1.781
Porosity (%)	0.12	0.39	0.59	0.71
Elastic modulus (GPa)	40.2 ± 1.3	42.2 ± 3.7	42.4 ± 0.5	43.0 ± 1.6
0.2% YS (MPa)	134 ± 4	123 ± 15	119 ± 5	123 ± 4
UTS (MPa)	199 ± 1	199 ± 13	197 ± 1	198 ± 6
Ductility (%)	5.9 ± 1.9	6.4 ± 1.1	6.4 ± 0.4	9.0 ± 2.1
WoF (J/m^3)	10.6 ± 3.9	10.4 ± 2.1	10.8 ± 0.2	15.8 ± 3.2
Macrohardness (HR15T)	46.9 ± 0.8	62.2 ± 3.0	63.0 ± 4.2	63.5 ± 3.7
Microhardness (HV)—matrix	38.9 ± 0.6	39.3 ± 1.5	39.5 ± 1.5	39.9 ± 1.4
Microhardness (HV)—Mg/Mo interface	—	59.8 ± 2.4	60.2 ± 3.9	61.6 ± 5.4

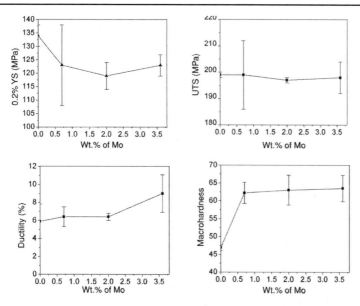

5.9.6. Addition of Aluminum

5.9.6.1. Mg Reinforced with Micrometer-Size Al (by Powder Metallurgy—Microwave Sintering)

Processing method	Powder metallurgy route
Sintering method	Hybrid microwave-assisted sintering (followed by water quenching for 5 min before extrusion)
Matrix material used	Pure Mg (98.5% purity)
Size of Mg used	60–300 μm
Size of Al used	7–15 μm
Sintering time	15 min
Extrusion temperature	350°C
Extrusion ratio	20.25:1

Refer to Table 5.70.

TABLE 5.70. Characteristics of Mg and Mg/7%Al [83].

	Material	Mg	Mg/7% Al
	Vol.% of Al	—	7.0
Tensile	0.2% Yield strength (MPa)	115.8 ± 10.3	130.2 ± 10.8
	Ultimate strength (MPa)	168.4 ± 9.8	170.4 ± 11.9
	Failure strain (%)	6.1 ± 2.0	4.0 ± 0.9
Compressive	0.2% Yield strength (MPa)	91.6 ± 11.5	152.5 ± 5.7
	Ultimate strength (MPa)	265.3 ± 5.3	286.1 ± 12.4
	Failure strain (%)	12.3 ± 0.6	7.3 ± 0.2
	Macrohardness (HR15T)	42.7 ± 1.8	50.8 ± 1.7

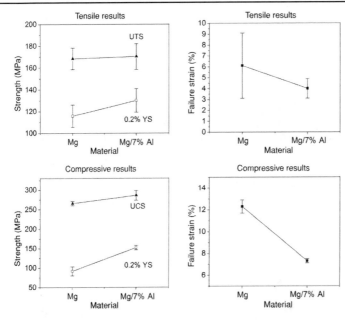

5.9.6.2. Mg Reinforced with 18 nm Al (by Powder Metallurgy—Conventional Sintering)

Processing method	Powder metallurgy route
Sintering method	Conventional tube furnace sintering
Matrix material used	Pure Mg (98.5% purity)
Size of Mg used	60–300 μm
Size of Al used	18 nm
Extrusion temperature	350°C
Extrusion ratio	25:1

Refer to Table 5.71.

TABLE 5.71. Characteristics of Mg and Mg/Al [84].

Material	Mg	Mg/0.38 Al	Mg/0.76 Al	Mg/1.16 Al	Mg/1.52 Al
Wt% of Al (vol.%)	—	0.38 (0.25)	0.76 (0.5)	1.16 (0.75)	1.52 (1.00)
Density (g/cm^3)	1.731	1.738	1.746	1.751	1.755
Porosity (%)	0.5	0.2	0.4	0.0	0.0
0.2% YS (MPa)	134 ± 11	181 ± 14	218 ± 16	202 ± 7	185 ± 9
UTS (MPa)	190 ± 10	221 ± 15	271 ± 11	261 ± 10	226 ± 12
Failure strain (%)	4.6 ± 0.6	4.8 ± 0.4	6.2 ± 0.9	5.0 ± 1.6	3.3 ± 1.0
WoF (MJ/m^3)	7.6 ± 1.6	10.5 ± 1.9	15.9 ± 2.1	13.1 ± 2.9	7.9 ± 1.8
Macrohardness (HR15T)	46 ± 3	54 ± 1	57 ± 1	60 ± 1	61 ± 1

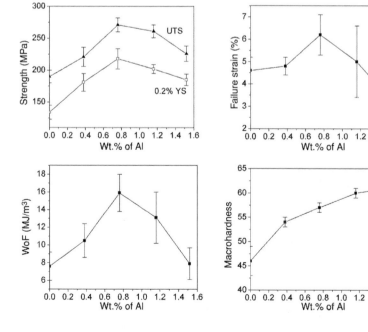

5.9.7. Addition of Iron Wire Mesh

5.9.7.1. Mg Reinforced with Interconnected Fe Wire Mesh (by Disintegrated Melt Deposition)

Processing method	Disintegrated melt deposition
Matrix material used	Pure Mg (>99.9% purity)
Interconnected reinforcement used	Galvanized Fe wire mesh
Extrusion temperature	350°C
Extrusion ratio	20.25:1

Refer to Table 5.72.

TABLE 5.72. Characteristics of Mg and Mg/Fe-wire composites [85].

	(1)	(2)	(3)	(4)
Material	Mg	Mg/4 star	Mg/6 star	Mg/8 star
Geometry of Fe wire[a]	—	(a) 4 star	(b) 6 star	(c) 8 star
Vol.% of Fe wire	—	1.77	2.65	3.58
Density (g/cm^3)	1.73	1.83	1.88	1.94
Porosity (%)	0.42	1.20	1.03	0.98
CTE ($\times 10^{-6}$/°C)	28.7 ± 0.4	28.4 ± 0.4	28.2 ± 0.3	28.1 ± 0.5
Dynamic modulus (GPa)	40.0	42.9	43.3	45.2
0.2% YS (MPa)	135 ± 11	153 ± 6	157 ± 11	163 ± 8
UTS (MPa)	233 ± 14	188 ± 11	190 ± 7	201 ± 8
Failure strain (%)	9.5 ± 2.3	4.1 ± 0.6	2.9 ± 0.6	2.5 ± 0.8
Macrohardness (HR15T)	39.0 ± 0.5	53.0 ± 0.6	53.2 ± 0.8	54.3 ± 0.8
Microhardness (HV)—matrix	37.2 ± 1.3	45.2 ± 2.3	47.4 ± 2.5	48.6 ± 2.7
Microhardness (HV)—Mg/Fe wire interface	—	207.6 ± 7.5	208.9 ± 9.3	208.3 ± 9.0

[a]Refer to schematic diagrams (a), (b), and (c) for the respective geometries of the Fe wire mesh.

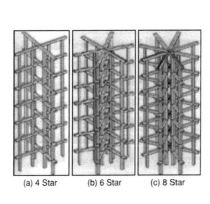

(a) 4 Star   (b) 6 Star   (c) 8 Star

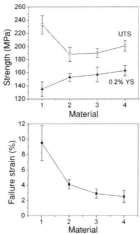

5.9.7.2. Mg Reinforced with Fe Wire Mesh and Carbon Fibers (by Disintegrated Melt Deposition)

Processing method	Disintegrated melt deposition
Matrix material used	Pure Mg (>99.9% purity)
Hybrid reinforcement used	Galvanized Fe wire mesh with PAN-based carbon fibers wound around the mesh
Extrusion temperature	350°C
Extrusion ratio	~20:1

Refer to Table 5.73.

TABLE 5.73. Characteristics of Mg and Mg/Fe-wire/CF composite [86].

	(1)	(2)
Material	Mg	Mg/Fe-wire/CF
Wt% of Fe wire (vol.%)	—	12.0 (3.0)
Wt% of carbon fiber (vol.%)	—	0.066 (0.071)
Density (g/cm^3)	1.738	1.896
Porosity (%)	0.12	1.20
CTE ($\times 10^{-6}$/°C)	29.1 ± 0.1	28.5 ± 0.8
Dynamic modulus (GPa)	39.8	42.5
0.2% YS (MPa)	153 ± 8	173 ± 4
UTS (MPa)	228 ± 18	224 ± 25
Ductility (%)	9.0 ± 1.5	3.0 ± 1.4
Macrohardness (HR15T)	47 ± 1	63 ± 5
Microhardness (HV)—Matrix	41 ± 1	51 ± 4
Microhardness (HV)—Mg/C fiber interface	—	78 ± 12
Microhardness (HV)—Mg/Fe wire interface	—	132 ± 30

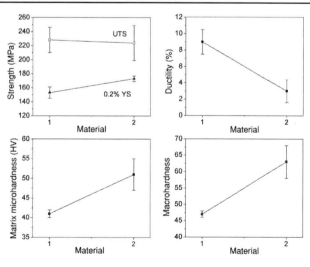

5.9.7.3. Mg reinforced with Fe Wire Mesh of Different Geometries (by Conventional Casting)

Processing method   Conventional casting
Matrix material used   Pure Mg (99.9% purity)
Reinforcement used   Galvanized Fe wire (0.9 mm diameter) with
                         10.8 vol.% Zn coating
Extrusion temperature   350°C
Extrusion ratio   20.25:1

Refer to Table 5.74.

TABLE 5.74. Characteristics of Mg and Mg/Fe-wire composite [87].

Material	(1) Mg	(2) Mg/Wire$_{spiral}$	(3) Mg/Wire$_{tubular}$
Geometry of Fe wire[a]	—	(a) Spiral	(b) Tubular
Vol. fraction of Fe wire	—	0.05	0.05
Density (g/cm^3)	1.740	2.026	2.042
Elastic modulus (GPa)	39 ± 2	47 ± 5	49 ± 2
0.2% YS (MPa)	140 ± 9	143 ± 22	141 ± 11
UTS (MPa)	218 ± 4	176 ± 13	185 ± 30
Ductility (%)	8.2 ± 1.1	2.0 ± 1.0	2.6 ± 1.8
Microhardness (HV)—matrix	38.3 ± 0.9	47.4 ± 1.7	47.5 ± 1.7

[a]Refer to schematic diagrams (a) and (b) for the respective geometries of the Fe wire.

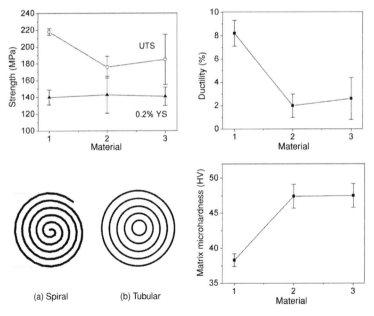

5.10. BIMETAL Mg/Al MACROCOMPOSITE

Macrocomposites are gaining importance as researchers are trying to further cut down the weight of automobiles by partially replacing aluminum-based alloys with magnesium alloys.

Extremely limited studies have been conducted to create macrocomposites based on magnesium and aluminum integration at millimeter length scale. Tensile and compressive behaviors of such composites are investigated and presented in this section.

The concept of development and the use of macrocomposites based on Mg–Al system is novel, but much effort and vision are required to use them in engineering applications.

5.10.1. Mg/Al Macrocomposite Containing Millimeter-Scale Al Core Reinforcement (by Disintegrated Melt Deposition)

Processing method	Disintegrated melt deposition
Matrix material used	Mg turnings (99.9+% purity)
Reinforcement used	Al lumps (99.5% purity)
Extrusion temperature	350°C
Extrusion ratio	20.25:1

(a) Tensile Test
Refer to Table 5.75.

TABLE 5.75. Tensile test results of Mg, Al, and Mg/Al macrocomposite [88].

Material	Mg	Al	Mg/0.079 Al
Volume Fraction of Al	—	1.000	0.079
CTE ($\times 10^{-6}$ /K)	29.0 ± 0.2	26.0 ± 0.6	27.0 ± 0.5
Elastic Modulus (GPa)	45.3	70.1	47.4
0.2% YS (MPa)	124 ± 11	137 ± 9	103 ± 10
UTS (MPa)	201 ± 13	183 ± 8	170 ± 10
Failure strain (%)	6.1 ± 1.1	19.4 ± 0.3	14.9 ± 2.9
WoF (MJ/m^3)	11.6 ± 2.6	32.6 ± 1.4	20.1 ± 4.4

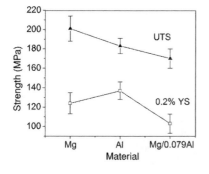
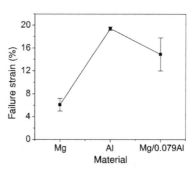

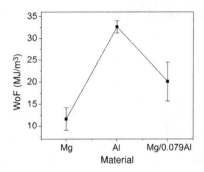

(b) Compression Test
Refer to Table 5.76.

TABLE 5.76. Compression test results of Mg and Mg/Al macrocomposite [89].

Material	Mg	Mg/0.081 Al	Mg/0.157 Al	Mg/0.223 Al
Volume fraction of Al	—	0.081	0.157	0.223
0.2% CYS (MPa)	74 ± 4	72	73	86
UCS (MPa)	353 ± 12	343	325	311
Failure strain (%)—compressive	16 ± 1	32	37	31
WOF (MJ/m^3)—compressive	43 ± 4	76	67	56

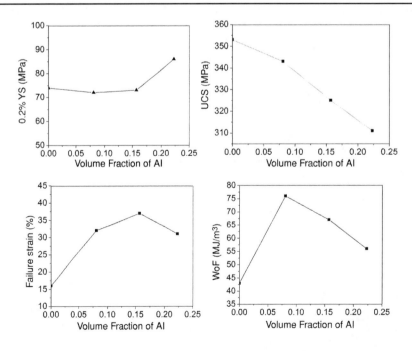

5.10.2. AZ31/AA5052 Macrocomposite Containing Millimeter-Scale Al Core Reinforcement (by Disintegrated Melt Deposition)

Processing method	Disintegrated melt deposition
Matrix material used	AZ31 rod (nominally 3 wt% Al, 1 wt% Zn, 0.2 wt% Mn, and balance Mg)
Reinforcement used	AA5052 (nominally 2.2–2.8 wt% Mg, 0.25 wt% Si, 0.40 wt% Fe, 0.10 wt% Cu, 0.10 wt% Mn, 0.15–0.35 wt% Cr, 0.10 wt% Zn, 0.15 wt% others, and balance Al)
Extrusion temperature	350°C
Extrusion ratio	20.25:1

Refer to Table 5.77.

TABLE 5.77. Compression test results of AZ31 and AZ31/0.079 AA5052 macrocomposite [90].

Material	AZ31	AZ31/0.079 AA5052
0.2% CYS (MPa)	93 ± 9	140 ± 11
UCS (MPa)	486 ± 4	505 ± 5
Failure strain (%)—compressive	19.7 ± 7.2	23.3 ± 5.3
WoF (MJ/m^3)—compressive	76 ± 14	114 ± 16

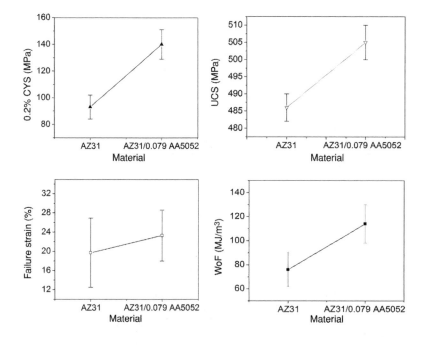

5.10.3. AZ31-Al₂O₃/AA5052 Macrocomposite Containing 50 nm Al₂O₃ and |Millimeter-Scale Al Core Reinforcement (by Disintegrated Melt Deposition)

Processing method	Disintegrated melt deposition
Matrix material used	AZ31 rod (nominally 2.50–3.50 wt% Al, 0.60–1.40 wt% Zn, 0.15–0.40 wt% Mn, 0.10 wt% Si, 0.05 wt% Cu, 0.01 wt% Fe, 0.01 wt% Ni, and balance Mg)
Reinforcement used	Al₂O₃ (50 nm) AA5052 (nominally 2.2–2.8 wt% Mg, 0.25 wt% Si, 0.40 wt% Fe, 0.10 wt% Cu, 0.10 wt% Mn, 0.15–0.35 wt% Cr, 0.10 wt% Zn, 0.15 wt% others, and balance Al)
Extrusion temperature	350°C
Extrusion ratio	20.25:1

(a) Tensile Test
Refer to Table 5.78.

TABLE 5.78. Tensile test results of AZ31 and AZ31–1.5 vol.% Al₂O₃/0.079 AA5052 macrocomposite [91].

Material	AZ31	AZ31–1.5 vol.% Al₂O₃/0.079 AA5052
Elastic modulus (GPa)	44 ± 1	61 ± 1
0.2% TYS (MPa)	172 ± 15	188 ± 10
UTS (MPa)	263 ± 12	312 ± 9
Failure strain (%)—tensile	10.4 ± 3.9	11.1 ± 0.4
WOF (MJ/m³)—tensile	26 ± 9	33 ± 2

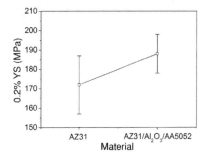
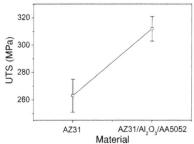

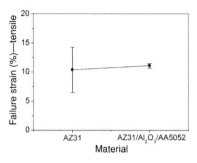 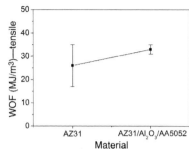

(b) Compression Test
Refer to Table 5.79.

TABLE 5.79. Compressive test results of AZ31 and AZ31–1.5 vol.% Al$_2$O$_3$/0.079 AA5052 macrocomposite [91].

Material	AZ31	AZ31–1.5 vol.% Al$_2$O$_3$/0.079 AA5052
0.2% CYS (MPa)	93 ± 9	147 ± 6
UCS (MPa)	486 ± 4	504 ± 10
Failure strain (%)—compressive	19.7 ± 7.2	21.9 ± 3.6
WOF (MJ/m^3)—compressive	76 ± 14	115 ± 15

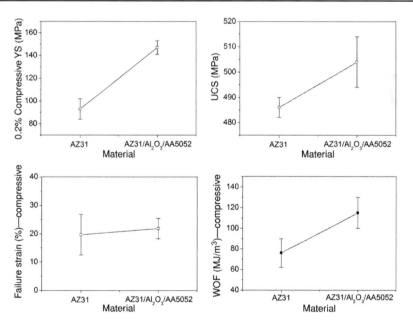

5.10.4. AZ31-CNT/AA5052 Macrocomposite Containing CNTs and Millimeter-Scale Al Core Reinforcement (by Disintegrated Melt Deposition)

Processing method	Disintegrated melt deposition
Matrix material used	AZ31 rod (nominally 3 wt% Al, 1 wt% Zn, 0.2 wt% Mn, and balance Mg)
Reinforcement used	Carbon nanotube (CNT); purity (94.7%); outer diameter (40 to 70 nm)
	AA5052 (nominally 2.2–2.8 wt% Mg, 0.25 wt% Si, 0.40 wt% Fe, 0.10 wt% Cu, 0.10 wt% Mn, 0.15–0.35 wt% Cr, 0.10 wt% Zn, 0.15 wt% others, and balance Al)
Extrusion temperature	350°C
Extrusion ratio	20.25:1

(a) Tensile Test
Refer to Table 5.80.

TABLE 5.80. Tensile test results of AZ31 and AZ31–1.0 vol.% CNT/0.079 AA5052 macrocomposite [92].

Material	AZ31	AZ31–1.0 vol.% CNT/0.079 AA5052
Elastic Modulus (GPa)	44 ± 1	61 ± 1
0.2% TYS (MPa)	172 ± 15	169 ± 6
UTS (MPa)	263 ± 12	296 ± 8
Failure strain (%)—tensile	10.4 ± 3.9	12.2 ± 0.8
WOF (MJ/m^3)—tensile	26 ± 9	33 ± 3

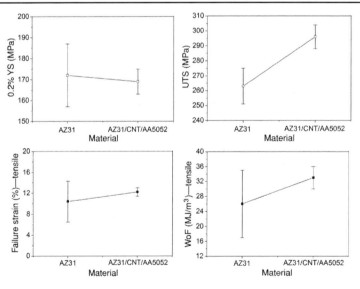

(b) Compression Test
Refer to Table 5.81.

TABLE 5.81. Compressive test results of AZ31 and AZ31–1.0 vol.% CNT/0.079 AA5052 macrocomposite [92].

Material	AZ31	AZ31–1.0 vol.% CNT/0.079 AA5052
0.2% CYS (MPa)	93 ± 9	126 ± 14
UCS (MPa)	486 ± 4	491 ± 7
Failure strain (%)—compressive	19.7 ± 7.2	27.9 ± 2.4
WOF (MJ/m^3)—compressive	76 ± 14	129 ± 18

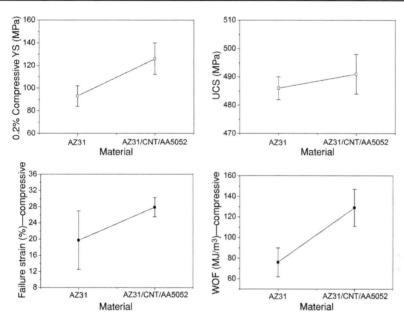

5.11. SUMMARY

This chapter presents the characteristics of a variety of magnesium-based composites. It is evident that the end properties of the magnesium-based composite can be tailored to suit specific requirements, with the judicious selection of type of matrix material (pure magnesium or magnesium alloy), type of reinforcement, length scale of reinforcement, amount of reinforcement, and also the processing method. In essence, this chapter serves as a useful guide for engineers, scientists, technicians, teachers, and students in the fields of materials design, development and selection, manufacturing, and engineering.

REFERENCES

1. D. J. Lloyd (1994) Particle reinforced aluminum and magnesium matrix composites. *International Materials Reviews*, **39**(1), 1–23.

2. H. Z. Ye and X. Y. Liu (2004) Review of recent studies in magnesium matrix composites. *Journal of Materials Science*, **39**(20), 6153–6171.

3. I. A. Ibrahim, F. A. Mohamed, and E. J. Lavernia. Particulate reinforced metal matrix composites—a review. *Journal of Materials Science*, **26**(5), 1137–1156.

4. M. Gupta and W. L. E. Wong (2007) *Microwaves and Metals*. Singapore: John Wiley & Sons, Ltd.

5. M. Gupta and W. L. E. Wong (2005) Enhancing overall mechanical performance of metallic materials using two-directional microwave assisted rapid sintering. *Scripta Materialia*, **52**(6), 479–483.

6. M. H. Nai, S. M. L. Nai, and M. Gupta (2008) Development and characterization of magnesium composites using nano-size oxide-based reinforcement. International Conference on Processing Materials for Properties (PMP-III), 7–10 Dec 2008, Bangkok, Thailand.

7. S. F. Hassan and M. Gupta (2004) Development of high-performance magnesium nano-composites using solidification processing route. *Materials Science and Technology*, **20**, 1383–1388.

8. S. F. Hassan and M. Gupta (2005) Development of high-performance magnesium nano-composites using nano-Al_2O_3 as reinforcement. *Materials Science and Engineering A*, **392**, 163–168.

9. S. F. Hassan and M. Gupta (2006) Effect of type of primary processing on the microstructure, CTE and mechanical properties of magnesium/alumina nanocomposites. *Composite Structures*, **72**, 19–26.

10. S. F. Hassan and M. Gupta (2006) Effect of length scale of Al_2O_3 particulates on microstructural and tensile properties of elemental Mg. *Materials Science and Engineering A*, **425**(1–2), 22–27.

11. J. C. Wong, M. Paramsothy, and M. Gupta (2009) Using Mg and Mg–nanoAl_2O_3 concentric alternating macro-ring material design to enhance the properties of magnesium. *Composites Science and Technology*, **69**(3–4), 438–444.

12. M. Paramsothy, S. F. Hassan, N. Srikanth, and M. Gupta (2009) Enhancing tensile/compressive response of magnesium alloy AZ31 by integrating with Al_2O_3 nanoparticles. *Materials Science and Engineering A*, **527**(1–2), 162–168.

13. Q. B. Nguyen and M. Gupta (2008) Increasing significantly the failure strain and work of fracture of solidification processed AZ31B using nano-Al_2O_3 particulates. *Journal of Alloys and Compounds*, **459**, 244–250.

14. Q. B. Nguyen and M. Gupta (2009) Microstructure and mechanical characteristics of AZ31B/Al_2O_3 nanocomposite with addition of Ca. *Journal of Composite Materials*, **43**(1), 5–17.

15. W. L. E. Wong, S. Karthik, and M. Gupta (2005) Development of hybrid Mg/Al$_2$O$_3$ composites with improved properties using microwave assisted rapid sintering route. *Journal of Materials Science*, **40**(13), 3395–3402.

16. S. K. Thakur, T. S. Srivatsan, and M. Gupta (2007) Synthesis and mechanical behavior of carbon nanotube–magnesium composites hybridized with nanoparticles of alumina. *Materials Science and Engineering A*, **466**(1–2), 32–37.

17. S. F. Hassan and M. Gupta (2008) Effect of submicron size Al$_2$O$_3$ particulates on microstructural and tensile properties of elemental Mg. *Journal of Alloys and Compounds*, **457**, 244–250.

18. W. L. E. Wong, S. Karthik, and M. Gupta (2005) Development of high performance Mg-Al$_2$O$_3$ composites containing Al$_2$O$_3$ in submicron length scale using microwave assisted rapid sintering. *Materials Science and Technology*, **21**(9), 1063–1070.

19. W. L. E. Wong, S. Karthik, and M. Gupta (2005) Development of hybrid Mg/Al$_2$O$_3$ composites with improved properties using microwave assisted rapid sintering route. *Journal of Materials Science*, **40**(13), 3395–3402.

20. W. L. E. Wong and M. Gupta (2007) Improving overall mechanical performance of magnesium using nano-alumina reinforcement and energy efficient microwave assisted processing route. *Advanced Engineering Materials*, **9**(10), 902–909.

21. X. L. Zhong and M. Gupta (2005) Effect of presence of nano-size alumina particles on the properties of elemental magnesium. *Journal of Metastable and Nanocrystalline Materials*, **23**, 171–174.

22. S. F. Hassan and M. Gupta (2006) Effect of particulate size of Al$_2$O$_3$ reinforcement on microstructure and mechanical behavior of solidification processed elemental Mg. *Journal of Alloys and Compounds*, **419**(1–2), 84–90.

23. S. F. Hassan, M. J. Tan, and M. Gupta (2008) High-temperature tensile properties of Mg/Al$_2$O$_3$ nanocomposite. *Materials Science and Engineering A*, **486**, 56–62.

24. Y. V. R. K. Prasad, K. P. Rao, and M. Gupta (2009) Hot workability and deformation mechanisms in Mg/nano-Al$_2$O$_3$ composite. *Composites Science and Technology*, **69**(7–8), 1070–1076.

25. Q. B. Nguyen, M. Gupta, and T. S. Srivatsan (2009) On the role of nano-alumina particulate reinforcements in enhancing the oxidation resistance of magnesium alloy AZ31B. *Materials Science and Engineering A*, **500**, 233–237.

26. Q. B. Nguyen and M. Gupta (2008) Enhancing compressive response of AZ31B magnesium alloy using alumina nanoparticulates. *Composites Science and Technology*, **68**(10–11), 2185–2192.

27. N. Srikanth, X. L. Zhong, and M. Gupta (2005) Enhancing damping of pure magnesium using nano-size alumina particulates. *Materials Letters*, **59**(29–30), 3851–3855.

28. Z. Trojanova, Z. Drozd, S. Kudela, Z. Szaraz, and P. Lukac (2007) Strengthening in Mg–Li matrix composites. *Composites Science and Technology*, **67**, 1965–1973.

29. C. S. Goh, M. Gupta, J. Wei, and L. C. Lee (2007) Characterization of high performance Mg/MgO nanocomposites. *Journal of Composite Materials*, **41**(19), 2325–2335.

30. P. Poddar, V. C. Srivastava, P. K. De, and K. L. Sahoo (2007) Processing and mechanical properties of SiC reinforced cast magnesium matrix composites by stir casting process. *Materials Science and Engineering A*, **460–461**, 357–364.

31. S. M. Kumar and B. K. Dhindaw (2007) Magnesium alloy—SiC$_p$ reinforced infiltrated cast composites. *Materials and Manufacturing Processes*, **22**, 429–432.

32. R. A. Saravanan and M. K. Surappa (2000) Fabrication and characterisation of pure magnesium-30 vol.% SiC$_P$ particle composite. *Materials Science and Engineering A*, **276**, 108–116.

33. S. C. V. Lim and M. Gupta (2001) Enhancing the microstructural and mechanical response of a Mg/SiC formulation by the method of reducing extrusion temperature. *Materials Research Bulletin*, **36**(15), 2627–2636.

34. S. C. V. Lim and M. Gupta (2003) Enhancing modulus and ductility of Mg/SiC composite through judicious selection of extrusion temperature and heat treatment. *Materials Science and Technology*, **19**, 803–808.

35. L. P. Soon and M. Gupta (2002) Synthesis and recyclability of a magnesium based composite using an innovative disintegrated melt deposition technique. *Materials Science and Technology*, **18**, 92–98.

36. M. Gupta, L. Lu, M. O. Lai, and K. H. Lee (1999) Microstructure and mechanical properties of elemental and reinforced magnesium synthesized using a fluxless liquid-phase process. *Materials Research Bulletin*, **34**(8), 1201–1214.

37. M. Gupta, M. O. Lai, and D. Saravanaranganathan (2000) Synthesis, microstructure and properties characterization of disintegrated melt deposited Mg/SiC composites. *Journal of Materials Science*, **35**(9), 2155–2165.

38. M. Manoharan, S. C. V. Lim, and M. Gupta (2002) Application of a model for the work hardening behavior to Mg/SiC composites synthesized using a fluxless casting process. *Materials Science and Engineering A*, **333**(1–2), 243–249.

39. S. C. V. Lim, M. Gupta, and L. Lu (2001) Processing, microstructure, and properties of Mg–SiC composites synthesised using fluxless casting process. *Materials Science and Technology*, **17**(7), 823–832.

40. W. Xie, Y. Liu, D. S. Li, J. Zhang, Z. W. Zhang, and J. Bi (2007) Influence of sintering routes to the mechanical properties of magnesium alloy and its composites produced by PM technique. *Journal of Alloys and Compounds*, **431**(1–2), 162–166.

41. M. C. Gui, J. M. Han, and P. Y. Li (2004) Microstructure and mechanical properties of Mg–Al9Zn/SiCp composite produced by vacuum stir casting process. *Materials Science and Technology*, **20**, 765–771.

42. S. U. Reddy, N. Srikanth, M. Gupta, and S. K. Sinha (2004) Enhancing the properties of magnesium using SiC particulates in sub-micron length scale. *Advanced engineering materials*, **6**(12), 957–964.

43. S. Ugandhar, M. Gupta, and S. K. Sinha (2006) Enhancing strength and ductility of Mg/SiC composites using recrystallization heat treatment. *Composite Structures*, **72**(2), 266–272.

44. G. Cao, H. Konishi, and X. Li (2008) Mechanical properties and microstructure of Mg/SiC nanocomposites fabricated by ultrasonic cavitation based nanomanufacturing. *Journal of Manufacturing Science and Engineering*, **130**(3), 031105.

45. G. Cao, J. Kobliska, H. H. Konishi, and X. Li (2008) Tensile properties and microstructure of SiC nanoparticle-reinforced Mg-4Zn alloy fabricated by ultrasonic cavitation-based solidification processing. *Metallurgical and Materials Transactions A*, **39A**, 880–886.

46. G. Cao, H. Choi, H. Konishi, S. Kou, R. Lakes, and X. Li (2008) Mg–6Zn/1.5%SiC nanocomposites fabricated by ultrasonic cavitation-based solidification processing. *Journal of Materials Science*, **43**, 5521–5526.

47. W. L. E Wong and M. Gupta (2006) Simultaneously improving strength and ductility of magnesium using nano-size SiC particulates and microwaves. *Advanced Engineering Materials*, **8**(8), 735–740.

48. W. L. E Wong and M. Gupta (2006) Effect of hybrid length scales (micro + nano) of SiC reinforcement on the properties of magnesium. *Solid State Phenomena*, **111**, 91–94.

49. S. K. Thakur, K. Balasubramanian, and M. Gupta (2007) Microwave synthesis and characterization of magnesium based composites containing nanosized SiC and hybrid (SiC +Al$_2$O$_3$) reinforcements. *Transactions of the ASME*, **129**, 194–199.

50. S. K. Thakur, T. K. Gan, and M. Gupta (2007) Development and characterization of magnesium composites containing nano-sized silicon carbide and carbon nanotubes as hybrid reinforcements. *Journal of Materials Science*, **42**, 10040–10046.

51. V. Kevorkijan (2003) AZ80 and ZC71/SiC/12p closed die forgings for automotive applications: technical and economic assessment of possible mass production. *Materials Science and Technology*, **19**, 1386–1390.

52. S. Tiwari, R. Balasubramaniam, and M. Gupta (2007) Corrosion behavior of SiC reinforced magnesium composites. *Corrosion Science*, **49**(2), 711–725.

53. C. Y. H. Lim, S. C. Lim, and M. Gupta (2003) Wear behaviour of SiCp-reinforced magnesium matrix composites. *Wear*, **255**(1–6), 629–637.

54. M. K. K. Oo, P. S. Ling, and M. Gupta (2000) Characteristics of Mg-based composites synthesized using a novel mechanical disintegration and deposition technique. *Metallurgical and Materials Transactions A*, **31**(7), 1873–1881.

55. S. F. Hassan and M. Gupta (2007) Development of nano-Y$_2$O$_3$ containing magnesium nanocomposites using solidification processing. *Journal of Alloys and Compounds*, **429**, 176–183.

56. S. F. Hassan and M. Gupta (2007) Development and characterization of ductile Mg/Y$_2$O$_3$ nanocomposites. *Transactions of the ASME*, **129**, 462–467.

57. C. S. Goh, J. Wei, L. C. Lee, and M. Gupta (2007) Properties and deformation behavior of Mg-Y$_2$O$_3$ nanocomposites. *Acta Materialia*, **55**(15), 5115–5121.

58. K. S. Tun and M. Gupta (2007) Improving mechanical properties of magnesium using nano-yttria reinforcement and microwave assisted powder metallurgy method. *Composites Science and Technology*, **67**(13), 2657–2664.

59. K. S. Tun and M. Gupta (2008) Effect of heating rate during hybrid microwave sintering on the tensile properties of magnesium and Mg/Y_2O_3 nanocomposite. *Journal of Alloys and Compounds*, **466**, 140–145.

60. K. S. Tun and M. Gupta (2008) Effect of extrusion ratio on microstructure and mechanical properties of microwave-sintered magnesium and Mg/Y_2O_3 nanocomposite. *Journal of Materials Science*, **43**, 4503–4511.

61. K. S. Tun, T. S. Srivatsan, and M. Gupta (2010) Investigating influence of hybrid (yttria+copper) nanoparticulate reinforcements on microstructural development and tensile response of magnesium. *Materials Science and Technology*, **26**(1), 87–94.

62. K. S. Tun and M. Gupta (2009) Development of magnesium/(yttria + nickel) hybrid nanocomposites using hybrid microwave sintering: microstructure and tensile properties. *Journal of Alloys and Compounds*, **487**(1–2), 76–82.

63. S. F. Hassan and M. Gupta (2007) Effect of Nano-ZrO_2 particulates reinforcement on microstructure and mechanical behavior of solidification processed elemental Mg. *Journal of Composite Materials*, **41**(21), 2533–2543.

64. S. F. Hassan, M. J. Tan, and M. Gupta (2007) Development of nano-ZrO_2 reinforced magnesium nanocomposites with significantly improved ductility. *Materials Science and Technology*, **23**(11), 1309–1312.

65. C. S. Goh, J. Wei, L. C. Lee, and M. Gupta (2006) Effect of fabrication techniques on the properties of carbon nanotubes reinforced magnesium. *Solid State Phenomena*, **111**, 179–182.

66. C. S. Goh, J. Wei, L. C. Lee, and M. Gupta (2006) Simultaneous enhancement in strength and ductility by reinforcing magnesium with carbon nanotubes. *Materials Science and Engineering A*, **423**, 153–156.

67. C. S. Goh, J. Wei, L. C. Lee, and M. Gupta (2006) Development of novel carbon nanotube reinforced magnesium nanocomposites using the powder metallurgy technique. *Nanotechnology*, **17**, 7–12.

68. Y. Shimizu, S. Miki, T. Soga, I. Itoh, H. Todoroki, T. Hosono, K. Sakaki, T. Hayashi, Y. A. Kim, M. Endo, S. Morimoto, and A. Koide (2008) Multi-walled carbon nanotube-reinforced magnesium alloy composites. *Scripta Materialia*, **58**, 267–270.

69. Q. Li, A. Viereckl, C. A. Rottmair, and R. F. Singer (2009) Improved processing of carbon nanotube/magnesium alloy composites. *Composites Science and Technology*, **69**, 1193–1199.

70. M. Paramsothy, S. F. Hassan, N. Srikanth, and M. Gupta (2010) Simultaneous enhancement of tensile/compressive strength and ductility of magnesium alloy AZ31 using carbon nanotubes. *Journal of Nanoscience and Nanotechnology*, **10**(2), 956–964.

71. C. S. Goh, J. Wei, L. C. Lee, and M. Gupta (2008) Ductility improvement and fatigue studies in Mg-CNT nanocomposites. *Composites Science and Technology*, **68**(6), 1432–1439.

72. S. F. Hassan and M. Gupta (2002) Development of a novel magnesium–copper based composite with improved mechanical properties. *Materials Research Bulletin*, **37**(2), 377–389.

73. S. F. Hassan and M. Gupta (2003) Development of high strength magnesium–copper based hybrid composites with enhanced tensile properties. *Materials Science and Technology*, **19**, 253–259.

74. K. F. Ho, M. Gupta, and T. S. Srivatsan (2004) The mechanical behavior of magnesium alloy AZ91 reinforced with fine copper particulates. *Materials Science and Engineering A*, **369**(1–2), 302–308.

75. S. F. Hassan, K. F. Ho, and M. Gupta (2004) Increasing elastic modulus, strength and CTE of AZ91 by reinforcing pure magnesium with elemental copper. *Materials Letters*, **58**(16), 2143–2146.

76. W. L. E. Wong and M. Gupta (2007) Development of Mg/Cu nanocomposites using microwave assisted rapid sintering. *Composites Science and Technology*, **67**(7–8), 1541–1552.

77. S. F. Hassan and M. Gupta (2002) Development of high strength magnesium based composites using elemental nickel particulates as reinforcement. *Journal of Materials Science*, **37**, 2467–2474.

78. S. F. Hassan and M. Gupta (2002) Development of a novel magnesium/nickel composite with improved mechanical properties. *Journal of Alloys and Compounds*, **335**, L10–L15.

79. S. F. Hassan and M. Gupta (2002) Development of ductile magnesium composite materials using titanium as reinforcement. *Journal of Alloys and Compounds*, **345**, 246–251.

80. Y. L. Xi, D. L. Chai, W. X. Zhang, and J. E. Zhou (2005) Ti–6Al–4V particle reinforced magnesium matrix composite by powder metallurgy. *Materials Letters*, **59**, 1831–1835.

81. M. A. Matin, L. Lu, and M. Gupta (2001) Investigation of the reactions between boron and titanium compounds with magnesium. *Scripta Materialia*, **45**(4), 479–486.

82. W. L. E. Wong and M. Gupta (2005) Enhancing thermal stability, modulus and ductility of magnesium using molybdenum as reinforcement. *Advanced Engineering Materials*, **7**(4), 250–256.

83. S. K. Thakur, M. Paramsothy, and M. Gupta (2010) Improving tensile and compressive strengths of magnesium by blending it with aluminium. *Materials Science and Technology*, **26**(1), 115–120.

84. X. L. Zhong, W. L. E. Wong, and M. Gupta (2007) Enhancing strength and ductility of magnesium by integrating it with aluminum nanoparticles. *Acta Materialia*, **55**(18), 6338–6344.

85. S. K. Thakur and M. Gupta (2008) Use of interconnected reinforcement in magnesium for stiffness critical applications. *Materials Science and Technology*, **24**(2), 213–220.

86. W. L. E. Wong and M. Gupta (2005) Using hybrid reinforcement methodology to enhance overall mechanical performance of pure magnesium. *Journal of Materials Science*, **40**, 2875–2882.

87. V. V. Ganesh and M. Gupta (2000) Synthesis and characterization of stiffness-critical materials using interconnected wires as reinforcement. *Materials Research Bulletin*, **35**, 2275–2286.

88. M. Paramsothy, N. Srikanth, and M. Gupta (2008) Solidification processed Mg/Al bimetal macrocomposite: microstructure and mechanical properties. *Journal of Alloys and Compounds*, **461**(1–2), 200–208.

89. M. Paramsothy, M. Gupta, and N. Srikanth (2008) Improving compressive failure strain and work of fracture of magnesium by integrating it with millimeter length scale aluminum. *Journal of Composite Materials*, **42**, 1297–1307.

90. M. Paramsothy, S. F. Hassan, N. Srikanth, and M. Gupta (2009) Enhancement of compressive strength and failure strain in AZ31 magnesium alloy. *Journal of Alloys and Compounds*, **482**, 73–80.

91. M. Paramsothy, S. F. Hassan, N. Srikanth, and M. Gupta (2009) Simultaneously Enhanced Tensile and Compressive Response of AZ31-NanoAl$_2$O$_3$-AA5052 Macrocomposite. *Journal of Materials Science*, **44**(18), 4860–4873.

92. M. Paramsothy, S. F. Hassan, N. Srikanth, and M. Gupta (2009) Adding carbon nanotubes and integrating with AA5052 aluminium alloy core to simultaneously enhance stiffness, strength and failure strain of AZ31 magnesium alloy. *Composites Part A: Applied Science and Manufacturing*, **40**(9), 1490–1500.

93. S. Iijima (1991) Helical microtubules of graphitic carbon. *Nature*, **354**, 56–58.

94. S. F. Hassan and M. Gupta (2006) Effect of different types of nano-size oxide particulates on microstructural and mechanical properties of elemental Mg. *Journal of Materials Science*, **41**, 2229–2236.

95. N. Eustathopoulos, M. G. Nicholas, and B. Drevet (1999) *Wettability at High Temperatures*, Vol. **3**. Pergamon Materials Series, Elsevier, UK, p. 198.

96. J. D. Gilchrist (1989) *Extraction Metallurgy*, 3rd edition. Great Britain: Pergamon Press, p. 148.

97. B. Q. Han and D. C. Dunand (2000) Microstructure and mechanical properties of magnesium containing high volume fractions of yttria dispersoids. *Materials Science and Engineering A*, **277**, 297–304.

98. H. Jynge and K. Motzfeldt (1980) Reactions between molten magnesium and refractory oxides. *Electrochimica Acta*, **25**, 139–143.

99. P. M. Kelly (1972) The effect of particle shape on dispersion hardening. *Scripta Metallurgica*, **6**(8), 647–656.

100. G. N. Hassold, E. A. Holm, and D. J. Srolovitz (1990) Effects of particle size on inhibited grain growth. *Scripta Metallurgica et Material*, **24**, 101–106.

101. E. Flores, J. M. Cabrera, and J. M. Prado (2004) Effect of clustering of precipitates on grain growth. *Metallurgical and Materials Transactions A*, **35**(13), 1097–1103.

102. W. Yang and W. B. Lee (1993) *Mesoplasticity and Its Applications. Materials Research and Engineering.* Germany: Springer-Verlag.

6

CORROSION ASPECTS OF MAGNESIUM-BASED MATERIALS

Magnesium-based materials are susceptible to corrosion particularly in wet atmospheric conditions. This may be principally attributed to the low electrode potential exhibited by magnesium. Over the years, effects have been made to enhance corrosion resistance of magnesium through judicious use of alloying elements and development of surface protection techniques. It is most important to note that all metals corrode in wet atmospheric conditions and magnesium alloy such as AZ31 exhibits comparable or superior corrosion resistance to 0.27% C steel in industrial, marine, and rural atmospheric conditions. This illustrates that provided efforts are made to improve surface protection techniques; magnesium-based materials can comfortably replace aluminum-based materials in many engineering applications in near future.

6.1. INTRODUCTION

Magnesium, magnesium alloys, and magnesium composites are strong candidates for existing and emerging engineering applications due to their low density and high strength-to-weight ratio. In dry atmospheric conditions, magnesium-based materials are quite stable from corrosion perspective due to the reasonably protective nature of their oxide

Magnesium, Magnesium Alloys, & Magnesium Composites, by Manoj Gupta and Nai Mui Ling, Sharon
© 2010 John Wiley & Sons, Inc.

TABLE 6.1. The standard emf series [1, 2].

	Electrode	Electrode Reaction	Standard Electrode Potential (V)
↑ Increasing active (Anodic)	Li, Li$^+$	Li$^+$ + e$^-$ → Li	−3.02
	K, K$^+$	K$^+$ + e$^-$ → K	−2.92
	Na, Na$^+$	Na$^+$ + e$^-$ → Na	−2.71
	Mg, Mg^{2+}	**Mg^{2+} + 2e$^-$ → Mg**	**−2.37**
	Al, Al^{3+}	Al^{3+} + 3e$^-$ → Al	−1.66
	Zn, Zn^{2+}	Zn^{2+} + 2e$^-$ → Zn	−0.76
	Cr, Cr^{3+}	Cr^{3+} + 3e$^-$ → Cr	−0.74
	Fe, Fe^{2+}	Fe^{2+} + 2e$^-$ → Fe	−0.44
	Cd, Cd^{2+}	Cd^{2+} + 2e$^-$ → Cd	−0.40
	Co, Co^{2+}	Co^{2+} + 2e$^-$ → Co	−0.28
	Ni, Ni^{2+}	Ni^{2+} + 2e$^-$ → Ni	−0.25
	Sn, Sn^{2+}	Sn^{2+} + 2e$^-$ → Sn	−0.14
	Pb, Pb^{2+}	Pb^{2+} + 2e$^-$ → Pb	−0.13
	H$_2$, H$^+$	**2H$^+$ + 2e$^-$ → H$_2$**	**0.00**
Increasing inert (Cathodic)	Cu, Cu^{2+}	Cu^{2+} + 2e$^-$ → Cu	0.34
	Ag, Ag$^+$	Ag$^+$ + e$^-$ → Ag	0.80
	Pt, Pt^{2+}	Pt^{2+} + 2e$^-$ → Pt	1.20
↓	Au, Au^{3+}	Au^{3+} + 3e$^-$ → Au	1.42

despite a Pilling–Bedworth ratio of 0.81. However, in aqueous environment, magnesium is anodic to most other metals due to its strongly negative electrode potential (see Table 6.1). Owing to its poor corrosion properties, magnesium-based materials are prone to galvanic corrosion, which can result in severe pitting. This, in turn, degrades the mechanical performance and greatly limits the usage of magnesium-based materials. Thus, understanding and enhancing the corrosion performance of magnesium-based materials is crucial to their future applications.

In this chapter, the key types of corrosion suffered by magnesium-based materials, the factors affecting corrosion and ways to improve the corrosion performance of magnesium-based materials are presented.

Table 6.1 lists the standard single-electrode potentials, which constitute the electromotive force series, also known as the emf series. It represents the corrosion tendencies of metals. The metals at the top of the table (such as lithium and potassium) are the most active. Those metals down the list are increasingly more inert (such as gold and platinum).

When magnesium is exposed to the atmosphere under room temperature condition, a surface film of magnesium oxide is formed. The presence of moisture in the atmosphere leads to the conversion of magnesium oxide into magnesium hydroxide [1]. In dilute NaCl solutions, the corrosion potential of magnesium is −1.73 V$_{nhe}$ (see Table 6.2). The formation of a Mg(OH)$_2$ or MgO surface film contributes to the difference in value between the standard potential and the actual corrosion potential [4–6].

TABLE 6.2. Corrosion potential of commonly used metals and alloys in 3–6% NaCl solutions [3].

Metal or Alloy	Corrosion Potential (V_{nhe})
Mg	**−1.73**
Mg alloys	**−1.67**
Mild steel, Zn plated	−1.14
Zn	−1.05
Mild steel, Cd plated	−0.86
Al (99.99%)	−0.85
Al 12% Si	−0.83
Mild steel	−0.78
Cast iron	−0.73
Pb	−0.55
Sn	−0.50
Stainless steel 316, in active state	−0.43
Brass (60/40)	−0.33
Cu	−0.22
Ni	−0.14
Stainless steel, in passive state	−0.13
Ag	−0.05
Au	0.18

A more useful and practical ranking to reflect the reactivity of metals is provided by the galvanic series as presented in Table 6.3. This series lists the relative reactivity of commercial metals and alloys in unpolluted seawater under room temperature condition. Moving down the list from No. 1 to No. 89, the commercial metal or alloys become cathodic and less reactive. Magnesium, which tops the list in the galvanic series, is the most anodic and most reactive metal. On the other extreme end, platinum is the most cathodic and most noble.

Comparing magnesium with aluminum and low carbon steels in terms of corrosion rate, it was found that magnesium alloy AZ31 exhibits superior corrosion performance than 0.27% C steel and inferior corrosion resistance than 2024 aluminum alloy in marine and rural atmospheric conditions. All the alloys were tested in sheet form over a period of 2.5 years.

6.2. TYPES OF CORROSION

Magnesium and its alloys are anodic and reactive. However, the formation of protective film on magnesium can aid to protect magnesium. As reported, even for exposures to marine atmospheres, the corrosion resistance of magnesium alloys is superior to that of mild steel [4, 8]. In hydrofluoric acid, magnesium alloys are capable to resist

TABLE 6.3. Galvanic series for commercial metals and alloys in seawater at room temperature [7].

No.	Metal and Alloys	No.	Metal
1	Magnesium	46	Yellow brass, 268
2	Mg alloy, AZ31B	47	Uranium 8% Mo
3	Mg alloy, HK31A	48	Naval brass, 464
4	Mg alloy (hot-dip, die cast or plated)	49	Muntz metal, 280
5	Beryllium (hot pressed)	50	Brass (plated)
6	Al 7072 clad on 7075	51	Nickel–silver (18% Ni)
7	Al 2014-T3	52	Stainless steel 316L (in active state)
8	Al 1160-H14	53	Bronze, 220
9	Al 7079-T6	54	Copper, 110
10	Cadmium (plated)	55	Red brass
11	Uranium	56	Stainless steel 347 (in active state)
12	Al 218 (die cast)	57	Molybdenum
13	Al 5052–0	58	Copper–nickel, 715
14	Al 5052-H12	59	Admiralty brass
15	Al 5456–0, H353	60	Stainless steel 202 (in active state)
16	Al 5052-H32	61	Bronze, phosphor, 534 (B-1)
17	Al 1100–0	62	Monte, 1 400
18	Al 3003-H25	63	Stainless steel 201 (in active state)
19	Al 6061-T6	64	Stainless steel 321 (in active state)
20	Al A360 (die cast)	65	Stainless steel 316 (in active state)
21	Al 7075-T6	66	Stainless steel 309 (in active state)
22	Al 6061-0	67	Stainless steel 17–7PH (in passive state)
23	Indium	68	Silicone bronze, 655
24	Al 2014-0	69	Stainless steel 304 (in passive state)
25	Al 2024-T4	70	Stainless steel 301 (in passive state)
26	Al 5052-H16	71	Stainless steel 321 (in passive state)
27	Tin (plated)	72	Stainless steel 201 (in passive state)
28	Stainless steel 430 (in active state)	73	Stainless steel 286 (in passive state)
29	Lead	74	Stainless steel 316L (in passive state)
30	Steel, 1010	75	AM 355 (in active state)
31	Iron (cast)	76	Stainless steel 202 (in passive state)
32	Stainless steel 410 (in active state)	77	AM355 (in passive state)
33	Copper (plated, cast, or wrought)	78	A286 (in passive state)
34	Nickel (plated)	79	Titanium 5 Al, 2.5 Sn
35	Chromium (plated)	80	Titanium 13 V, 11 Cr, 3 Al (annealed)
36	Tantalum	81	Titanium 6 Al, 4 V (solution treated and aged)
37	AM350 (in active state)	82	Titanium 6 Al, 4 V (annealed)
38	Stainless steel 310 (in active state)	83	Titanium 8 Mn
39	Stainless steel 301 (in active state)	84	Titanium 13 V, 11 Cr, 3 Al (solution treated and aged)
40	Stainless steel 304 (in active state)		
41	Stainless steel 430 (in active state)	85	Titanium 75 A
42	Stainless steel 410 (in active state)	86	AM350 (in passive state)
43	Stainless steel 17–7PH (in active state)	87	Silver
44	Tungsten	88	Gold
45	Niobium (columbium) 1% Zr	89	Platinum

corrosion because of the formation of magnesium fluoride, which is highly insoluble [9]. Furthermore, under clean and dry conditions, magnesium alloys are practically inert for indefinite periods.

Two key factors that contribute to the poor corrosion resistance of magnesium alloys are as follows [10]:

(a) The presence of secondary phases or impurities resulting in internal galvanic corrosion [11].
(b) The formation of quasi-passive hydroxide film on magnesium, which is less stable than those passive films formed on stainless steels and aluminum.

The main types of corrosion experienced by magnesium-based materials are [1,3,9,12] as follows:

(a) Galvanic corrosion
(b) Pitting corrosion
(c) Intergranular corrosion
(d) Stress corrosion
(e) High temperature oxidation

6.2.1. Galvanic Corrosion

Magnesium is a highly reactive metal due to its low standard electrode potential (-2.37 V, see Table 6.1). It is anodic (or sacrificial) to all other engineering metals. It is susceptible to corrosion attack, as it readily forms galvanic corrosion system due to

(a) coupling with another metal,
(b) presence of secondary phases (such as α phase and/or β phase), and
(c) presence of impurities.

There are two types of galvanic corrosion (see Figure 6.1):

(a) Internal galvanic corrosion due to the presence of secondary phases or impurities.
(b) External galvanic corrosion due to coupling with other engineering materials.

The extent of corrosion in a given environment is dependent on the relative positions of the two metals (galvanic couple) in the electrochemical series (see Table 6.1). In order to reduce galvanic corrosion, there must be adequate surface protection (such as coatings), elimination of contaminations, and proper assembly design with judicious coupling with other materials.

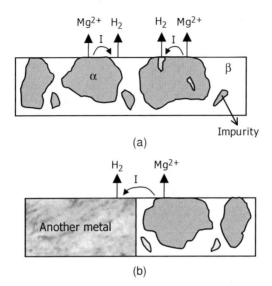

Figure 6.1. Types of galvanic corrosion: (a) internal and (b) external galvanic corrosion.

6.2.2. Pitting Corrosion

Pitting corrosion is a form of localized corrosion, whereby it results in the formation of pits (holes) in the corroded metal. In the case of Mg–Al alloys, selective attack along the $Mg_{17}Al_{12}$ network results in pitting. Air contaminated with salt also tends to break down the protective film leading to formation of pits.

6.2.3. Intergranular Corrosion

Magnesium and magnesium alloys are practically not prone to intergranular corrosion because the grain boundary constituent is usually cathodic to the grain body [13]. However, corrosion is intense in the grains and the areas adjoining the grain boundary.

6.2.4. Stress Corrosion Cracking

Commercially pure magnesium is not prone to stress corrosion cracking (SCC) when loaded up to its yield strength under most environments. The presence of zinc and aluminum as alloying additions promotes SCC susceptibility in magnesium alloys [14, 15]. Wrought and cast magnesium alloys, especially aluminum-containing magnesium alloys, have the highest tendency toward SCC [1]. In most cases, SCC in magnesium is transgranular. Transgranular SCC is mainly triggered by hydrogen embrittlement [9]. However, when alloys are subjected to heat treatments, the change in material's grain size and the precipitation of phases at grain boundaries can result in intergranular SCC. In the case of Mg–Al–Zn alloys, intergranular SCC takes place due to the precipitation of $Mg_{17}Al_{12}$ along the grain boundaries [4].

Zirconium-containing magnesium alloys are reported to be free from SCC except under stresses approaching yield stress [11]. Magnesium alloys free from both aluminum and zinc additions are most resistant to SCC. Mg–Mn alloys are one of such alloys that are inert to SCC when loaded up to the yield strength under normal environments [1].

6.2.5. High Temperature Oxidation

When magnesium is exposed to oxygen, magnesium oxide film is formed on the surface. This oxide film provides a good protection under dry oxygen environment below 450°C and under moist oxygen environment under 380°C [9]. With increasing oxidation temperature, it was reported that the magnesium oxide film changes into a porous structure and is no longer protective [16, 17].

Furthermore, with the addition of alloying elements such as aluminum and zinc, the rate of oxidation increases accordingly. On the other hand, the addition of minute amounts of lanthanum and cerium has reported to aid in reducing the rate of oxidation, to a level below that of pure magnesium [18]. You et al. [19] also reported the formation of MgO/CaO protective layer on the surface of Mg–Ca alloys at elevated temperature. This dense and compact layer assists to retard oxidation of magnesium at high temperature.

6.3. INFLUENCE OF IMPURITY

The presence of different elemental impurity has different effects on the corrosion performance of magnesium materials. Some elements are known to be detrimental to the corrosion resistance, while some have insignificant or unknown effects. In order to minimize the corrosion of magnesium materials, high purity magnesium alloys with heavy metal impurities (such as iron, copper, and nickel) below a certain threshold value should be used. Fe, Cu, and Ni are harmful due to their low solid solubility limits. There exists a tolerance limit, whereby the corrosion rate will significantly increase when the concentration of impurity exceeds the limit (see Figure 6.2). When their concentrations exceed their respective tolerance limit, they segregate and serve as active cathodic sites for electrochemical attack.

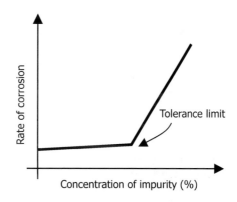

Figure 6.2. Graphical illustration showing the effect of concentration of impurity (Fe, Ni, and Cu) on the rate of corrosion of magnesium [3].

In a study by Hanawalt et al. [19], it was reported that Fe, Ni, Cu, and Co, even at concentrations of less than 0.2%, had very strong accelerating effect on the saltwater corrosion rate. Studies by other researchers [20, 21] have also documented that the purity of metal is the most vital factor in the corrosion behavior.

6.4. CORROSION BEHAVIOR OF MAGNESIUM-BASED MATERIALS

6.4.1. Addition of Rare Earth Element

Rare earth elements are often used as alloying elements in magnesium. In a study by Rosalbino et al. [22], Er was incorporated into a Mg–Al alloy and it was reported that the presence of Er in $Mg(OH)_2$ lattice is accountable for the enhanced corrosion resistance of the Mg–Al–Er alloy.

Zhou et al. [23] investigated the effect of holmium addition in the AZ91D magnesium alloy. With the addition of 0.24 wt% and 0.44 wt% Ho in the Mg–9Al alloy, the corrosion resistance of the alloy was improved significantly as shown in Figure 6.3. This was attributed to the following:

(a) The optimized microstructure of the Mg alloy with the addition of Ho. The concentration of Fe in the intermetallic phases and the fraction of β phase in the alloys were reduced with the incorporation of Ho. As the β phase in Mg–9Al alloys functions as an active cathode [24], the presence of lower amount of β

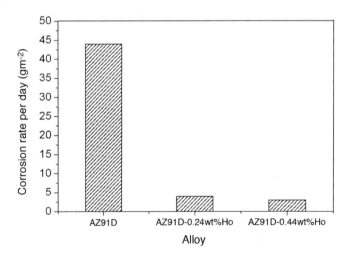

Figure 6.3. Graphical illustration showing the average corrosion rates of AZ91D alloy and AZ91D alloy with Ho additions in 3.5 wt% NaCl solution at 25°C, measured by sample weight loss for 24 h [23].

phase in the alloys retarded the microgalvanic corrosion caused by its coupling with anodic Mg-based α phase.

(b) The formation of more compact, smoother and uniform films on the surface of Ho-containing alloys, thus rendering them more protective against corrosion.

6.4.2. Addition of Reinforcements

Magnesium-based composites have gained increasing interest in various industries for applications, which need high performance materials with lightweight characteristics. Magnesium-based composites, in general, exhibit poor corrosion resistance because of their reactive nature [25]. Furthermore, the presence of reinforcements in the magnesium matrix aid to increase its corrosion sensitivity through the following ways [25]:

(a) Galvanic coupling between the reinforcement and the magnesium-based matrix.
(b) Selective corrosion at the interface due to formation of interfacial phase between the matrix and reinforcement.
(c) Microstructural changes resulting from the synthesis of composites.
(d) Addition of processing contaminants resulting from the synthesis of composites.

6.4.2.1. Addition of Metallic Reinforcements.
In a study by Budruk Abhijeet et al. [26], the corrosion behavior of Mg–Cu (0.3, 0.6, and 1.0 vol.%) and Mg–Mo (0.1, 0.3, and 0.6 vol.%) composites were investigated in 3.5% NaCl solution using the weight loss and polarization method. Corrosion results revealed that as the volume fraction of reinforcement in the Mg–Cu and Mg–Mo composites increased, the corrosion rate increased correspondingly. Both composites exhibited higher corrosion rates than that of pure magnesium. Furthermore, with the same amount of reinforcement addition, magnesium composites containing molybdenum corroded faster than that reinforced with copper. The poor corrosion resistance of the composites was attributed to

(a) microgalvanic influences between the magnesium matrix and the reinforcements, and
(b) surface films' lower quality.

6.4.2.2. Addition of SiC.
In a study by Pardo et al. [27], the corrosion behavior of AZ92 magnesium alloy reinforced with silicon carbide (SiC) particulates in the following environments was investigated:

(a) 3.5 wt% NaCl solution
(b) Neutral salt fog (according to ASTM B 117)
(c) High relative humidity (98% RH, 50°C) environment

The corrosion results showed that the incorporation of SiC enhanced the corrosion rate of the composite materials after immersion in 3.5 wt% NaCl solution for 1 h

TABLE 6.4. Corrosion potential and corrosion rate of AZ92 and AZ92/SiC composites after immersion in 3.5 wt% NaCl solution for 1 h [27].

Material	E_{corr} (V/SSE)	Corrosion Rate ($\times 10^{-2}$) (mg cm^{-2} h^{-1})
AZ92	−1.457	4.2
AZ92/5 vol.% SiC	−1.472	4.4
AZ92/10 vol.% SiC	−1.480	5.6

(see Table 6.4). The presence of reinforcements also encouraged cracking and spalling of the corrosion layer for increasing exposure times. However, under the high humidity environment, the reinforcement had minimal effect on the corrosion behavior and the composites exhibited a higher corrosion resistance.

In another study by Hihara et al. [28, 29], the galvanic corrosion of pure Mg/SiC and ZE41A/SiC in following oxygenated and deaerated solutions was investigated:

(a) Nitrate (0.5 M NaNO$_3$) solution
(b) Sulphate (0.5 M Na$_2$SO$_4$) solution
(c) Chloride (3.15% NaCl) solution

The SiC reinforcement was in the form of a monofilament that consists of a carbon core and a carbon-rich surface. The results revealed high galvanic corrosion rates of the composites in the presence of oxygen. This was because the SiC monofilaments acted as effective O$_2$ reduction sites. Moreover, using ZE41A alloy as the matrix material yielded better corrosion performance than using pure Mg.

Tiwari et al. [30] also studied the corrosion behavior of pure magnesium reinforced with 6 vol.% and 16 vol.% of SiC particulates, in freely aerated 1 M NaCl solution. The results (see Table 6.5) evidently showed that SiC particulates degraded the corrosion resistance of magnesium, and with increasing reinforcement addition, the corrosion resistance decreased accordingly. It was suggested that the increase in corrosion rate was not due to the galvanic corrosion between the reinforcement and the matrix, but

TABLE 6.5. Corrosion results of Mg and Mg/SiC samples [30].

Material	Zero Current Potential (V vs. SSC)	Corrosion Rate (Tafel)[a] (mm/yr)
Mg	−1.116	13.5
Mg/6 vol.% SiC	−1.138	18.1
Mg/12 vol.% SiC	−1.145	45.5

[a]Determined using Tafel extrapolation.

rather due to the defective nature of the surface film as observed from the electrochemical impedance spectroscopy.

Xue and his coworkers [31] investigated the corrosion behavior of microarc oxidation (MAO) coatings on stir-cast AZ31/6 vol.% SiC_p magnesium composite. Due to the addition of reinforcements in magnesium composites, conventional anodizing and chemical conversion coatings are deemed less corrosion protective as the growth of the oxide film is retarded by the presence of the reinforcements and this, in turn, disrupts the coating's continuity [32]. Thus, the MAO treatment in $Na_3PO_4 + KOH + NaF$ electrolyte was used in this study to form a thick compact corrosion protective coating (up to 80 μm thick). It was reported that most of the SiC particulates were oxidized under the microarc discharge sintering, with some SiC particulates remaining in the coating. It was also observed that despite the presence of these residual SiC particulates, the continuity of the MAO coating was not disrupted. The uncoated and coated $AZ31/SiC_p$ samples were also subjected to electrochemical polarization test in 3.5 wt% NaCl solution. The results revealed that after MAO treatment, the corrosion current density decreased by approximately 3 orders of magnitude and the corrosion resistance was greatly enhanced. Furthermore, it was observed that the thickness of the coating influenced the corrosion resistance.

6.4.2.3. Addition of Al_2O_3.
Bakkar and his coinvestigators investigated the corrosion behavior of a magnesium alloy reinforced with alumina [33]. Firstly, the AS41 magnesium composite was fabricated using the Saffil-alumina (Al_2O_3) fiber preform via the squeeze casting technique. The samples were then subjected to electrochemical test and hydrogen evolution test to evaluate the corrosion characteristics of the AS41 alloy and its composite. The test results showed that the $AS41/Al_2O_3$ composite sample had comparable corrosion behavior to its unreinforced counterpart. With the addition of Al_2O_3 fibers, the corrosion potential did not change significantly.

Nguyen et al. [34] incorporated varying volume percentage of nanosize Al_2O_3 particulates (0.66 vol.%, 1.11 vol.%, and 1.50 vol.%) into AZ31B alloy to form Mg composites. The composites were tested under ambient atmospheric conditions, at temperatures ranging from 300 to 470°C, for 7 h. The results showed an increasing trend of oxidation resistance of the composite materials with the presence of increasing amount of Al_2O_3 particulates (see Figure 6.4).

6.4.2.4. Addition of Carbon Fibers.
Bakkar et al. [35] reinforced AS41 Mg alloy and AS41 (0.5 Ca) Mg alloy with short carbon fibers (Sigrafil C-40), using the squeeze casting method. The Mg alloy was forced into the short carbon fiber preform, where the carbon fibers (~25 vol.%) were quasi-isotropically distributed in the horizontal plane. The electrochemical test results revealed that with the addition of Ca in AS41 alloy, the corrosion resistance of the alloy was greatly improved. However, with the presence of carbon fibers in AS41 (0.5 Ca), the composite's corrosion resistance was compromised. Carbon is an electrically conductive element and is electrochemically unstable, though it is chemically stable in aqueous chloride environment. Furthermore, applying electrochemical polarization on the composites results in crevice corrosion at the C fiber/Mg interface. Under free immersion conditions, pitting corrosion was

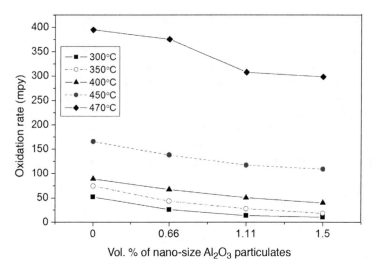

Figure 6.4. Graphical illustration showing the oxidation rate of AZ31B and its composites calculated using the Wagner's model at various temperatures [34].

dominating, however, the pitting sites showed no inclination to initiate at the C fiber/Mg interface. The galvanic coupling with C fibers results in severe corrosion of the Mg composites.

6.4.2.5. Addition of Multiwalled Carbon Nanotubes. A study by Endo et al. [36] reported significant improvement in the anticorrosive characteristic of AZ91D/MWCNT (multiwalled carbon nanotubes) composites. AZ91D alloy incorporated with 0, 1, and 5 wt% MWCNTs was immersed into saltwater (3 wt% NaCl, 293 K). After 20 h in the saltwater, results revealed that the unreinforced alloy was dissolved up to 13%; however, the alloys reinforced with 1 wt% MWCNT and 5 wt% MWCNT kept most of its initial weight below 2%. Furthermore, in the case of alloy containing 5 wt% MWCNTs, the corrosion current measurement study using tap water showed that after 4 h, the corrosion current totally stopped. On the contrary, for the case of unreinforced alloy, the corrosion current was virtually the same throughout the entire investigation period of 24 h.

This finding of high corrosion resistance behavior exhibited by magnesium alloy composites was attributed to the following:

(a) The hydrophobic characteristic of MWCNTs. With the presence of MWCNTs in the magnesium alloy, the water repellency of the composite alloy is enhanced. Hence, this improved the corrosion resistance of the composite alloy as corrosion takes place when there is excess amount of water in contact with the surface of the alloy.

(b) The formation of stable oxide film along the grain boundaries of the magnesium alloy composite. The presence of MWCNTs keeps the oxide layer from detaching from the alloy and this retards further formation of oxide layer. However, in the case of unreinforced alloy, more cracks were found on the surface that results in the continuous detachment of the oxide layer.

6.5. WAYS TO REDUCE CORROSION

In order to mitigate the corrosion of magnesium-based materials so as to improve their service performance and to increase their range of applications, it is essential to devise ways to reduce/prevent corrosion. The following lists some of the reported and proven corrosion prevention measures [3, 37]:

(a) Use of surface modification technology such as laser annealing and ion implantation.

(b) Application of protective coatings and films to provide a barrier between the environment and the magnesium-based material.

(c) Use of high purity alloys and decrease the amount of impurities below their allowable tolerance limits.

(d) Develop new magnesium-based materials with the addition of new alloying elements, reinforcements or secondary phases.

One of the proven and most effective ways to prevent corrosion is to coat the base material. However, in order to achieve effective corrosion protection, the coating film should

(a) adhere well to the base material,

(b) be free from pores,

(c) be uniform, and

(d) possess self-healing capability for applications where physical damage to the coating may occur.

There are many different technologies to coat magnesium-based materials and they can be broadly grouped into [3, 37]

(i) chemical treatments,

(ii) hydride coatings,

(iii) anodizing,

(iv) plasma electrolytic oxidation (PEO) treatment,

(v) conversion coatings,

(vi) thermal spray coatings, and

(vii) organic/polymer coatings.

6.5.1. Chemical Treatments

These include, for example, treatment of magnesium samples with chromate solutions. Such treatments clean and protect the material surface with the formation of $Mg(OH)_2$ film and a chromium compound. The protective film forms a good base for subsequent organic coatings on magnesium [4, 18].

Although these conventional anticorrosion coatings that contain chromates are effective, they are harmful to the environment [38, 39]. In view of this, efforts are also made to develop environmentally friendly coatings [37, 40]. Hydroxyapatite (HA) is one of such potential coatings. It possesses the following characteristics that made it a favorable choice of environmentally friendly anticorrosion coatings for magnesium alloys:

(a) It is not environmentally toxic, as it is a basic mineral component of human bone.

(b) Its structure is thermodynamically stable, thus will exhibit high corrosion resistance [41].

(c) The synthesis of HA is also environmentally friendly, as it does not contain any toxic substances [42, 43].

In a study reported by Hiromoto et al. [44], HA was directly synthesized on pure magnesium and AZ series alloys (AZ31, AZ61, and AZ91) using a Ca chelate compound. Corrosion results revealed that for the case of a HA-coated pure magnesium, no visible corrosion pits were seen after the cyclic dry and wet tests with 1 gm^{-2} NaCl on the surface and polarization tests in a 3.5 wt% NaCl solution. However, for the cases of HA-coated Mg alloy samples, after 8–24 h in a 3.5 wt% NaCl solution, 10^3–10^4 times lower anodic current density was registered than for the case of as-polished samples in the polarization test. These results evidently showed the great potential of the HA coating as an anticorrosion protection for pure magnesium and its alloys.

6.5.2. Hydride Coatings

These are formed through an electrochemical process that involves treating the magnesium substrate (cathode) with an alkaline solution that is prepared using alkali metal hydroxide or ammonium salts. First, the samples are mechanically polished, degreased with solvent, and then subjected to acid etch before cathodic treatment. Nakatsugawa et al. [45, 46] reported a 33.3% decrease in the corrosion rate of AZ91D alloy with the hydride coating compared with the dichromate treatment.

6.5.3. Anodizing

Anodizing treatments are used to produce a thick, stable oxide film on metals and alloys through an electrolytic passivation process [47]. The steps of anodizing treatment are as shown in Figure 6.5. These anodized films play a role to enhance the adhesion of paint to the metal. As the coatings are usually porous even when thick, a sealing process

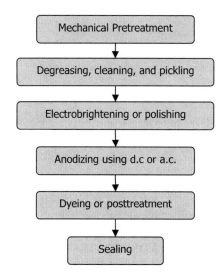

Figure 6.5. Steps of anodizing treatment.

(using steam treatment, lacquer sealing, dichromate sealing, or boiling in hot water) is required to achieve the necessary corrosion resistance [47, 48]. Hydrated base metal species are precipitated inside the pores to effectively seal off the oxide film.

The Dow Chemical Company developed Dow 17, the first anodized magnesium coating in the mid-1940s. The electrolyte solution for the coating was made up of sodium dichromate, ammonium bifluoride and phosphoric acid. It contained a high amount of chromium (VI) and was dark orange in color. It also had a pH of approximately 5 and operated at or above 160°F.

The HAE anodic coating is another widely used anodic coating developed by Harry A. Evangelides in mid-1950s. The coating is applied with AC and the electrolyte used was made up of potassium permanganate, potassium fluoride, trisodium phosphate, potassium hydroxide, and aluminum hydroxide. Due to the presence of permanganate, it was deep purple in color. It was also a very high alkaline solution and had a pH of approximately 14 and operated between 70°F and 86°F.

In view of the ever-stricter service conditions in today's defense, military, and aerospace industries, the Tagnite coating [49] was developed in 1992 to replace the HAE and the Dow 17 anodize coatings that did not possess the required protection properties. Furthermore, the key advantage over the HAE and Dow 17 electrolytes is that the Tagnite electrolyte is a clear, alkaline solution, which does not contain any chromium (IV) or heavy metals. It was developed keeping in mind the Clean Water and Clean Air Act initiative. It is made up of hydroxide, fluoride, and silicate. The solution is operated below room temperature at 40–60°F. The Tagnite coating has proven itself to be more corrosion resistance than Dow 17 and HAE anodize coatings. This can be attributed to the following:

(a) Its improved coating structure. The Tagnite coating consists of the buildup of tightly packed magnesium oxide.

TABLE 6.6. Potentiodynamic polarization test results of uncoated and coated AZ91D alloy in 3.5 wt% NaCl solution at 30 ± 1 °C [50].

Material	E_{corr} (mV)	i_{corr} (μA cm^{-2})
Uncoated AZ91D	−1535	38.360
AZ91D with new coating	−1379	0.184
AZ91D with Dow 17 coating	−1577	1.694
AZ91D with HAE coating	−1630	1.496

(b) The presence of smaller pores, which retards the penetration of corrosive medium through the coating, to reach the base magnesium metal.

The Tagnite coating has been extensively used in the power tool industry and also in the aerospace industry for the past decade. In the latter application, it is used to coat transmission cases, gearboxes, oil pump housings, and so on. The Tagnite coating also exhibits superior paint adhesion.

Zhang and his coworkers [50] presented a new, environmentally friendly coating for magnesium and its alloys. The electrolyte solution comprised potassium hydroxide, sodium carbonate, sodium silicate, and sodium borate. It was operated between 5°C and 85°C. The electrochemical test results of uncoated and coated AZ91D samples are presented in Table 6.6. Furthermore, after subjected to 48 h of total immersion test at 20 ± 1°C in 3.5 wt% NaCl solution (refer to Table 6.7), the results evidently revealed that the new coating process is more corrosion protective than HAE and Dow 17 coatings.

6.5.4. PEO Treatment

In recent years, PEO treatment is gaining popularity as one of the surface modification methods for magnesium materials [40,51,52]. PEO, which is also known as MAO, is an electrochemical surface treatment process that produces ceramic-like coatings on metals like magnesium, aluminum, and titanium with incorporation of species originating from

TABLE 6.7. Average corrosion rate of uncoated and coated AZ91D samples after 48 h of total immersion test [50].

Material	Film Thickness (μm)	Average Corrosion Rate (mg m^{-2} h^{-1})
Uncoated AZ91D	0	258
AZ91D with new coating	10	50
	17	42
	33	33
AZ91D with Dow 17 coating	31	222
AZ91D with HAE coating	19	149

TABLE 6.8. Potentiodynamic polarization test results of
uncoated and PEO-coated AZ91D alloys in 3.5 wt% NaCl
solution (pH 6.5) at room temperature [55].

Material	E_{corr} (V)	i_{corr} (A cm^{-2})
Uncoated AZ91D	−1.4838	4.7134×10^{-5}
PEO-coated AZ91D	−1.4326	1.8598×10^{-7}

the substrate and the electrolyte [53]. MAO and PEO evolved from the conventional anodizing process.

For the PEO treatment of magnesium alloys, silicate, phosphate, or aluminate-containing alkaline electrolytes with the coating containing amorphous or crystalline phases (such as MgO, Mg_2AlO_4, $Mg_3(PO_4)_2$, and Mg_2SiO_4) [40], are commonly used. The PEO coatings can be generated under the AC, DC, or bipolar electrical regimes [54].

Luo and his coworkers [55] successfully prepared a composite coating (Al_2O_3–ZrO_2–Y_2O_3) on AZ91D magnesium alloy using the PEO method in an alkaline aluminate electrolyte. The results revealed that the corrosion potential of the PEO-coated AZ91D alloy was improved and the corrosion current density was greatly decreased by 2 orders of magnitude when compared with that of the uncoated AZ91D (see Table 6.8) alloy. The lower I_{corr} value translates to better corrosion resistance behavior.

Arrabal et al. [56] investigated the AC PEO of magnesium alloys (AZ31, AZ61, AZ91D, ZC71, ZE41, and WE43-T6). From the potentiodynamic polarization tests in 3.5 wt% NaCl, the results showed that the corrosion rate of the magnesium alloys was reduced by 2–4 orders of magnitude after the PEO treatment.

In existing literature, there are limited reports on PEO coatings formed on composites. In another study, Arrabal and his coworkers [57] investigated the corrosion behavior of PEO of a magnesium composite (ZC71 reinforced with 12 vol.% SiC). It was observed that the SiC particles were incorporated into the coatings, and the PEO coating improved the corrosion resistance of the magnesium composite in 3.5 wt% NaCl, after immersion for 20 min in the solution.

Bala Srinivasan and his coworkers [58] applied a silicate-based DC PEO treatment on a cast AM50 magnesium alloy. The electrochemical test results showed that the PEO treatment enhanced the corrosion resistance of the AM50 alloy in 0.1 M NaCl solution (see Table 6.9).

TABLE 6.9. Electrochemical test results of uncoated
and PEO-coated AM50 alloys in 0.1 M NaCl solution [58].

Material	i_{corr} (mA cm^{-2})
Uncoated AM50	1.9×10^{-3}
PEO-coated AM50	1.2×10^{-5}

Cao et al. [59] reported the PEO of AM60 alloys using 50 Hz AC anodizing method in an alkaline borate solution with silicate additives. The AM60 alloys with the PEO film exhibited superior corrosion resistance. The polarization results revealed that the PEO coating aids to decrease the corrosion current by 3–4 orders of magnitude compared with that of the uncoated AM60 alloy.

6.5.5. Conversion Coatings

Conversion coatings are a form of surface treatments that are formed by chemical or electrochemical processes. The metal surface is converted into a layer of metal oxides, chromates, phosphates, or other compounds that are chemically bonded to the surface. As with all surface treatments, in order to achieve a good conversion coating, the sample needs to be properly cleaned and pretreated prior to coating. As chromate, phosphate/permanganate, and fluorozirconate treatments are commonly used to form the conversion coatings, the toxicity of these treatment solutions is a major concern. In recent years, there is an urgent move to search for more environmentally acceptable alternatives. This has resulted in the use of rare earth metal salts (REMS), in particular cerium species [60].

In a study by Rudd et al. [60], corrosion protection in the case of pure magnesium and WE43 magnesium alloy was achieved by the use of rare earth (cerium, lanthanum, and praseodymium) conversion coatings. The coatings were formed on the metal surface by immersion in the rare earth salt containing solutions. The electrochemical test results revealed that the samples of coated pure magnesium and coated WE43 alloy had improved corrosion resistance in a pH 8.5 buffer solution.

In another study by Ardelean et al. [61], an attempt was made to minimize corrosion of AZ91 and AM50 alloys with cerium-, zirconium-, and niobium-based conversion coatings. The alloys were treated in an aqueous solution of $Ce(NO_3)_3$, $ZrO(NO_3)_2$, and $Nb_xO_yF_z$, at pH 4 and room temperature, over a range of testing duration. The potentiodynamic polarization curves evidently exhibited the enhanced corrosion protection characteristics of the Ce–Zr–Nb-based coating in terms of

(a) decreased corrosion and anodic dissolution current densities,
(b) increased corrosion potential, and
(c) increased polarization resistance.

6.5.6. Thermal Spray Coatings

In this technique, first the base material to be coated should be cleaned and roughened before treatment. Then, polymeric, metal, ceramic, or cermet is introduced to a torch and then heated to a temperature near or above its melting point. The coating on the material is formed by accelerating the molten droplets through a gas stream, whereby allowing the droplets to adhere to the surface of the material [62–66].

This technique has the following benefits:

(a) Ability to use any coating material that melts without decomposing.
(b) During the deposition process, there is minimal heating of the material to be coated.
(c) Ability to strip and recoat, without affecting the properties or dimensions of the base material.

However, this technique also has its disadvantages:

(a) Lack of ability to coat cavities with high aspect ratio and surface area that is parallel to the direction of spraying.
(b) Sealing and mechanical finishing are required because of porous nature of coating.
(c) Health-, environment-, and safety-related issues due to the generation of fumes, dust, and light radiation during the process.

Campo et al. [64] investigated the corrosion behavior of thermally sprayed Al and Al/SiC composite coatings on pure magnesium. For the composite coatings, the pure aluminum powder (99.5 wt% Al, size $\sim$ 125 μm) was firstly homogeneously mixed with the SiC particulates (size $\sim$ 52 μm), using a ball milling machine assisted with alumina balls for 15 min. The composite mixture was then fed into the spray gun and thermally sprayed onto the pure magnesium samples. The as-sprayed samples were also compacted at room temperature for 1 minute under 314 MPa to achieve nearly pore-free, dense coatings. The electrochemical test results revealed that the coated samples exhibited improved corrosion potentials when compared with that of their uncoated counterparts (see Table 6.10).

The results also showed that the current densities of as-sprayed Al/SiC- and Al-coated Mg samples were approximately 2 orders of magnitude and 3 orders of magnitude lower, respectively, compared with that of their uncoated counterparts. In the case of the

TABLE 6.10. Polarization test results of uncoated, as-sprayed, and compacted coated pure magnesium samples in 3.5 wt% NaCl solution after 1 h of immersion [64].

Material	E_{corr} (V)
Uncoated pure Mg	−1.496
Al-coated Mg (as-sprayed)	−1.280
Al/SiC-coated Mg (as-sprayed)	−1.427
Al-coated Mg (compacted coating)	−1.243
Al/SiC-coated Mg (compacted coating)	−1.197

samples with compacted Al and Al/SiC coatings, the current densities were 5–6 orders of magnitude lower than that of the uncoated sample.

6.5.7. Organic/Polymer Coatings

Organic/polymer coatings are one of the surface coating technologies. There are a variety of organic coating processes [37], which include the following:

(a) Painting
(b) Sol–gel process
(c) Powder coating
(d) Plasma polymerization
(e) Cathodic epoxy electrocoating
(f) Polymer plating

The primers for magnesium alloys are based on resins such as acrylic, polyvinyl butyral, vinyl, polyurethane, epoxy, baked phenolic with zinc or strontium chromates, and titanium dioxide pigments [67–69]. However, owing to the high toxicity and carcinogenic characteristics of chromates, other inhibiting pigments such as zinc phosphate and zinc molybdate are used [70]. In recent years, literature has shown that conducting polymers such as polyaniline and polypyrrole offer good corrosion protection for aluminum and magnesium alloys [37, 71].

Sathiyanarayanan et al. [72] studied the corrosion-resistant properties of polyaniline–acrylic coating on magnesium alloy ZM21. The paint coating contains phosphate-doped polyaniline. Results showed that after 75 days of exposure to NaCl solution, the protective coating remains intact on the magnesium alloy and there is no base metal dissolution. In another related study [73], ZM21 alloy was coated with the polyaniline-pigmented paint based on an epoxy binder and the corrosion behavior of the coating on the alloy was investigated. The test results revealed that in comparison with the chromate-pigmented paint, the polyaniline-pigmented paint offered similar protection for the ZM21 alloy.

Polyoxadiazole-based coating is another group of potential polymer coating, which is also used as conducting protecting coatings, as reported by Kannan et al. [71]. Polymers containing oxadiazole is known for their high degradation temperature ($\sim$500°C) and high chemical and mechanical stability [74]. In this study, POD-DPE (poly(4,4'-diphenyl-ether-1,3,4-oxadiazole)) and POD-6FP (poly(4,4'-diphenyl-hexafluoropropane-1,3,4-oxadiazole)) were coated on AM50 magnesium alloy. Results from electrochemical tests showed that POD-6FP-coated samples had higher corrosion resistance (in terms of significantly lower corrosion current, i_{corr}) than those coated with POD-DPE. As seen in Table 6.11, the i_{corr} value of the POD-DPE-coated sample is comparable to that of the uncoated alloy. However, the i_{corr} value of POD-6FP (1.4×10^{-6} mA/cm^2) is about 3 orders of magnitude lower than that of POD-DPE-coated sample (1.2×10^{-3} mA/cm^2). This finding can be attributed to the POD-6FP coating having a hydrophobic group [75], which results in the attachment to the polyoxadiazole chain.

TABLE 6.11. Polarization test results of uncoated and coated AM50 magnesium alloy samples in 0.1 M NaCl solution [71].

Material	E_{corr} (mV)	i_{corr} (mA cm^{-2})
Uncoated AM50 alloy	−1378	5.6×10^{-3}
POD-6FP	−1364	1.4×10^{-6}
POD-DPE	−1485	1.2×10^{-3}

6.6. SUMMARY

Like all other metals, magnesium-based materials also corrode. Corrosion of magnesium-based materials is a concern specially in wet atmospheric conditions, and efforts are made to improve inherent corrosion resistance of magnesium through alloying and through surface protection. Among the above approaches, the alloying approach is an ideal way to address the problem but more challenging as the alloying elements do change other mechanical properties and at times unfavorably. Surface protection approach is more realistic as the tailored mechanical properties of a given alloy are not compromised. Scientists and engineers are working extremely hard to bring the surface protection technologies for magnesium-based materials close to what we have for iron-based materials.

REFERENCES

1. M. M. Avedesian and H. Baker (eds) (1999) *ASM Specialty Handbook—Magnesium and Magnesium Alloys*. Materials Park, OH: ASM International.
2. W. D. Callister (2003) *Materials Science and Engineering: An introduction*. New York: John Wiley & Sons, Ltd.
3. G. Song and A. Atrens (1999) Corrosion mechanisms of magnesium alloys. *Advanced Engineering Materials*, **1**(1), 11–33.
4. G. L. Makar and J. Kruger (1993) Corrosion of magnesium. *International Materials Review*, **38**(3), 138–153.
5. J. L. Robinson and P. F. King (1961) Electrochemical behavior of the magnesium anode. *Journal of the Electrochemical Society*, **108**(1), 36–41.
6. K. Huber (1953) Anodic formation of coatings on magnesium, zinc and cadmium. *Journal of the Electrochemical Society*, **100**(8), 376–382.
7. Z. Ahmad (2006) *Principles of Corrosion Engineering and Corrosion Control*. Oxford: Butterworth-Heinemann.
8. O. Lunder, J. E. Lein, T. K. Aune, and K. Nisancioglu (1989) The role of $Mg_{17}Al_{12}$ phase in the corrosion of magnesium alloy AZ91. *Corrosion*, **45**(9), 741–748.
9. J. Kruger (2008) Magnesium alloys: Corrosion. In K. H. J. Buschow, R. W. Cahn, M. C. Flemings, and B. Ilschner (eds) *Encyclopedia of Materials: Science and Technology*, pp. 4744–4745.

10. G. L. Makar and J. Kruger (1990) Corrosion studies of rapidly solidified magnesium alloys. *Journal of the Electrochemical Society*, **137**(2), 414–421.

11. E. F. Emley (1966) *Principles of Magnesium Technology*. New York: Pergamon Press.

12. R. C. Zeng, J. Zhang, W. J. Huang, W. Dietzel, K. U. Kainer, C. Blawert, and W. Ke (2006) Review of studies on corrosion of magnesium alloys. *Transactions of Nonferrous Metals Society of China*, **16**, s763–s771.

13. E. Ghali, W. Dietzel, and K. U. Kainer (2004) General and localized corrosion of magnesium alloys: a critical review. *Journal of Materials Engineering and Performance*, **13**(1), 7–23.

14. H. Hessing and H. L. Logan (1955) *Corr. Rev. Control*, **2**, 53.

15. H. L. Logan (1958) Mechanism of stress-corrosion cracking in the AZ31B magnesium alloy. *Journal of Research of the National Bureau of Standards*, **61**, 503–508.

16. S. Lu, X. Y. Zhou, J. Chen, and Z. D. Hou (2007) Corrosion and high-temperature oxidation of AM60 magnesium alloy. *Transactions of Nonferrous Metals Society of China*, **17**, s156–s160.

17. B. S. You, W. W. Park, and I. S. Chung (2000) The effect of calcium additions on the oxidation behavior in magnesium alloys. *Scripta Materialia*, **42**, 1089–1094.

18. W. S. Loose (1946) Corrosion and protection of magnesium. In L. M. Pidgeon, J. C. Mathes, and N. E. Woldmen (eds) *Metals Handbook*. Materials Park, OH: ASM International, pp. 173–260.

19. J. D. Hanawalt, C. E. Nelson, and J. A. Peloubet (1942) Corrosion studies of magnesium and its alloys. *Transaction of AIME*, **147**, 273–299.

20. R. W. Murray and J. E. Hillis (1990) Magnesium Finishing: Chemical Treatment and Coating Particles. SAE Technical Paper Series #900 791, Detroit.

21. J. E. Hillis. (1983) The Effects of Heavy Metal Contamination on Magnesium Corrosion Performance. SAE Technical Paper #830 523, Detroit.

22. F. Rosalbino, E. Angelini, S. De Negri, A. Saccone, and S. Delfino (2005) Effect of erbium addition on the corrosion behavior of Mg-Al alloys. *Intermetallics*, **13**(1), 55–60.

23. X. H. Zhou, Y. W. Huang, Z. L. Wei, Q. R. Chen, and F. X. Gan (2006) Improvement of corrosion resistance of AZ91D magnesium alloy by holmium addition. *Corrosion Science*, **48**(12), 4223–4233.

24. G. Song, A. Atrens, X. L. Wu, and B. Zhang (1998) Corrosion behavior of AZ21, AZ501 and AZ91 in sodium chloride. *Corrosion Science*, **40**(10), 1769–1791.

25. L. H. Hihara and R. M. Latanision (1994) Corrosion of metal matrix composites. *International Materials Review*, **39**(6), 245–264.

26. S. Budruk Abhijeet, R. Balasubramaniam, and M. Gupta (2008) Corrosion behavior of Mg-Cu and Mg-Mo composites in 3.5% NaCl. *Corrosion Science*, **50**, 2423–2428.

27. A. Pardo, S. Merino, M. C. Merino, I. Barroso, M. Mohedano, R. Arrabal, and F. Viejo (2009) Corrosion behaviour of silicon-carbide particle reinforced AZ92 magnesium alloy. *Corrosion Science*, **51**(4), 841–849.

28. L. Hihara and P. K. Kondepudi (1993) The galvanic corrosion of SiC monofilament/ZE41 Mg metal-matrix composite in 0.5M NaNO$_3$. *Corrosion Science*, **34**(11), 1761–1772.

29. L. Hihara and P. K. Kondepudi (1994) Galvanic corrosion between SiC monofilament and magnesium in NaCl, Na$_2$SO$_4$ and NaNO$_3$ solutions for application to metal-matrix composites. *Corrosion Science*, **36**(9), 1585–1595.

30. S. Tiwari, R. Balasubramaniam, and M. Gupta (2007) Corrosion behavior of SiC reinforced magnesium composites. *Corrosion Science*, **49**(2), 711–725.

31. W. B. Xue, Q. Jin, Q. Z. Zhu, M. Hua, and Y. Y. Ma (2009) Anti-corrosion microarc oxidation coatings on SiC$_p$/AZ31 magnesium matrix composite. *Journal of Alloys and Compound*, **482**(1–2), 208–212.

32. M. Shahid (1997) Mechanism of film growth during anodizing of Al-alloy-8090/SiC metal matrix composite in sulphuric acid electrolyte. *Journal of Materials Science*, **32**(14), 3775–3781.

33. A. Bakkar and V. Neubert (2007) Corrosion characterisation of alumina–magnesium metal matrix composites. *Corrosion Science*, **49**(3), 1110–1130.

34. Q. B. Nguyen, M. Gupta, and T. S. Srivatsan (2009) On the role of nano-alumina particulate reinforcements in enhancing the oxidation resistance of magnesium alloy AZ31B. *Materials Science and Engineering A*, **500**, 233–237.

35. A. Bakkar and V. Neubert (2009) Corrosion behavior of carbon fibres/magnesium metal matrix composite and electrochemical response of its constituents. *Electrochimica Acta*, **54**(5), 1597–1606.

36. M. Endo, T. Hayashi, I. Itoh, Y. A. Kim, D. Shimamoto, H. Muramatsu, Y. Shimizu, S. Morimoto, M. Terrones, S. Iinou, and S. Koide (2008) An anticorrosive magnesium/carbon nanotube composite. *Applied Physics Letters*, **92**, 063105.

37. J. E. Gray and B. Luan (2002) Protective coatings on magnesium and its alloys—critical review. *Journal of Alloys and Compounds*, **336**, 88–113.

38. P. Kurze (2006) Corrosion and surface protections. In H. E. Friedrich, B. L. Mordike (eds) *Magnesium Technology—Metallurgy, Design Data, Applications*. New York: Springer, pp. 431–468.

39. The Japan Institute of Light Metals (ed.) (2000) *Handbook of Advanced Magnesium Technology*. Tokyo: Kallos Publishing, Chapters 1, 8 and 11.

40. C. Blawert, W. Dietzel, E. Ghali, and G. Song (2006) Anodizing treatments for magnesium alloys and their effect on corrosion resistance in various environments. *Advanced Engineering Materials*, **8**, 511–533.

41. D. MacConnell (1973) *Apatite: Its Crystal Chemistry, Mineralogy, Utilization, and Geologic and Biologic Occurrences*. New York: Springer, Chapter 5.

42. A. Bigi, G. Falini, E. Foresti, A. Ripamonti, M. Gazzano, and N. Roveri (1993) Magnesium influence on hydroxyapatite crystallization. *Journal of Inorganic Biochemistry*, **49**(1), 69–78.

43. I. V. Fadeev, L. I. Shvorneva, S. M. Barinov, and V. P. Orlovskii (2003) Synthesis and structure of magnesium-substituted hydroxyapatite. *Inorganic Materials*, **39**(9), 947–950.

44. S. Hiromoto and A. Yamamoto (2009) High corrosion resistance of magnesium coated with hydroxyapatite directly synthesized in an aqueous solution. *Electrochimica Acta*, **54**(27), 7085–7093.

45. I. Nakatsugawa (1996) Surface modification technology for magnesium products. *International Magnesium Association*, p. 24.

46. I. Nakatsugawa (2000) *Cathodic protective coating on magnesium or its alloys and method of producing the same.* US Patent 6117298.

47. C. K. Mittal (1995) Transactions of the Metal Finishers Association of India, **4**, 227.

48. A. Brace (1997) *Transactions*, **75**, B101.

49. Technology Applications Group—Tagnite Coating. Website: http://www.tagnite. com/tagnite_coating/ (last accessed on July 20, 2008).

50. Y. J. Zhang, C. W. Yan, F. H. Wang, H. Y. Lou, and C. N. Cao (2002) Electrochemical behavior of anodized Mg alloy AZ91D in chloride containing aqueous solution. *Surface and Coatings Technology*, **161**, 36–43.

51. C. Blawert, V. Heitmann, W. Dietzel, H. M. Nykyforchyn, and M. D. Klapkiv (2005) Influence of process parameters on the corrosion properties of electrolytic conversion plasma coated magnesium alloys. *Surface and Coatings Technology*, **200**, 68–72.

52. Y. Ma, X. Nie, D. O. Northwood, and H. Hu (2004) Corrosion and erosion properties of silicate and phosphate coatings on magnesium. *Thin Solid Films*, **469/470**, 472–477.

53. R. Arrabal, E. Matykina, P. Skeldon, G. E. Thompson, and A. Pardo (2008) Transport of species during plasma electrolytic oxidation of WE43-T6 magnesium alloy. *Journal of the Electrochemical Society*, **155**(3), C101–C111.

54. S. Xin, L. Song, R. Zhao, and X. Hu (2006) Influence of cathodic current on composition, structure and properties of Al_2O_3 coatings on aluminum alloy prepared by micro-arc oxidation process. *Thin Solid Films*, **515**(1), 326–332.

55. H. Luo, Q. Cai, J. He, and B. Wei (2009) Preparation and properties of composite ceramic coating containing $Al_2O_3–ZrO_2–Y_2O_3$ on AZ91D magnesium alloy by plasma electrolytic oxidation. *Current Applied Physics*, **9**(6), 1341–1346.

56. R. Arrabal, E. Matykina, T. Hashimoto, P. Skeldon, and G. E. Thompson (2009) Characterization of AC PEO coatings on magnesium alloys. *Surface and Coatings Technology*, **203**(16), 2207–2220.

57. R. Arrabal, E. Matykina, P. Skeldon, and G. E. Thompson (2009) Coating formation by plasma electrolytic oxidation on ZC71/SiC/12p-T6 magnesium metal matrix composite. *Applied Surface Science*, **255**(9), 5071–5078.

58. P. Bala Srinivasan, C. Blawert, and W. Dietzel (2008) Effect of plasma electrolytic oxidation treatment on the corrosion and stress corrosion cracking behaviour of AM50 magnesium alloy. *Materials Science and Engineering A*, **494**, 401–406.

59. F. H. Cao, L. Y. Lin, Z. Zhang, J. Q. Zhang, and C. N. Cao (2008) Environmental friendly plasma electrolytic oxidation of AM60 magnesium alloy and its corrosion resistance. *Transactions of Nonferrous Metal Society of China*, **18**(2), 240–247.

60. A. L. Rudd, C. B. Breslin, and F. Mansfeld (2000) The corrosion protection afforded by rare earth conversion coatings applied to magnesium. *Corrosion Science*, **42**(2), 275–288.

61. H. Ardelean, I. Frateur, and P. Marcus (2008) Corrosion protection of magnesium alloys by cerium, zirconium and niobium-based conversion coatings. *Corrosion Science*, **50**(7), 1907–1918.

62. B. Gérard (2006) Application of thermal spraying in the automobile industry. *Surface and Coatings Technology*, **201**(5), 2028–2031.

63. R. H. Unger (1987) Thermal spray coatings. *ASM Handbook: Corrosion*, **13**, 459.

64. M. Campo, M. Carboneras, M. D. López, B. Torres, P. Rodrigo, E. Otero, and J. Rams (2009) Corrosion resistance of thermally sprayed Al and Al/SiC coatings on Mg. *Surface and Coatings Technology*, **203**(20–21), 3224–3230.

65. H. Pokhmurska, B. Wielage, T. Lampke, T. Grund, M. Student, and N. Chervinska (2008) Post-treatment of thermal spray coatings on magnesium. *Surface and Coatings Technology*, **202**(18), 4515–4524.

66. L. H. Chiu, C. C. Chen, and C. F. Yang (2005) Improvement of corrosion properties in an aluminum-sprayed AZ31 magnesium alloy by a post-hot pressing and anodizing treatment. *Surface and Coatings Technology*, **191**(2–3), 181–187.

67. J. E. Hillis (1994) Surface engineering of magnesium alloys. *ASM Handbook on Surface Engineering*, Vol. **5**, ASTM International, 819 pp.

68. C. R. Hegedus, D. F. Pulley, S. J. Spadafora, A. T. Eng, and D. J. Hirst (1989) review of organic coating technology for U.S. naval aircraft. *Journal of Coatings Technology*, **61**(778), 31–44.

69. W. Wilson (1970) Method of Coating Magnesium Metal to Prevent Corrosion. US Patent 3537879.

70. P. J. Kinlen, V. Menon, and Y. W. Ding (1999) A mechanistic investigation of polyaniline corrosion protection using the scanning reference electrode technique. *Journal of the Electrochemical Society*, **146**(10), 3690–3695.

71. M. Bobby Kannan, D. Gomes, W. Dietzel, and V. Abetz (2008) Polyoxadiazole-based coating for corrosion protection of magnesium alloy. *Surface Coatings and Technology*, **202**(19), 4598–4601.

72. S. Sathiyanarayanan, S. Syed Azim, and G. Venkatachari (2006) Corrosion resistant properties of polyaniline-acrylic coating on magnesium alloy. *Applied Surface Science*, **253**(4), 2113–2117.

73. S. Sathiyanarayanan, S. Syed Azim, and G. Venkatachari (2008) Corrosion protection of magnesium alloy ZM21 by polyaniline-blended coatings. *Journal of Coatings Technology and Research*, **5**(4), 471–477.

74. H. H. Yang (1989) *Aromatic High-strength Fibers*. New York: John Wiley & Sons, Ltd.

75. E. David, A. Lazar, and A. Armeanu (2004) Surface modification of polytetrafluoroethylene for adhesive bonding. *Journal of Materials Processing Technology*, **157–158**, 284–289.

7

STRENGTH–DUCTILITY COMBINATIONS OF MAGNESIUM-BASED MATERIALS

This chapter allows the engineers, scientists, technicians, teachers, and students in the fields of materials design, development and selection, manufacturing and engineering to effectively select the magnesium system (with the desired strength and ductility) that best suits their application. For easy referencing, the yield strengths of the magnesium-based materials (monolithic/unreinforced materials and composite materials) are ranked in ascending order in each category, with their corresponding ductility values.

7.1. 0.2% YIELD STRENGTH < 100 MPa AND DUCTILITY MATRIX

(a) Monolithic/Unreinforced Materials

Material	Processing	0.2% Yield Strength (MPa)	Ductility (%)	Reference
K1A-F (tested at 315°C)	Sand casting	14	78 (EL)	[1]
AM100A-T6 (tested at 260°C)	Sand casting	28	45 (EL)	[1]
K1A-F (tested at 200°C)	Sand casting	34	71 (EL)	[1]

Magnesium, Magnesium Alloys, & Magnesium Composites, by Manoj Gupta and Nai Mui Ling, Sharon
© 2010 John Wiley & Sons, Inc.

Material	Processing	0.2% Yield Strength (MPa)	Ductility (%)	Reference
ZK51A-T5 (tested at 315°C)	Sand casting	41	16 (EL)	[1]
Mg4Zn	Melt casting—ultrasonic cavitation	42	8.5	[2]
AM100A-T6 (tested at 200°C)	Sand casting	45	25 (EL)	[1]
K1A-F (tested at 93°C)	Sand casting	48	30 (EL)	[1]
K1A-F	Sand & permanent mould casting	51	20 (EL)	[3]
Mg6Zn	Melt casting—ultrasonic cavitation	54	7.7 (EL)	[4]
EZ33A-T5 (tested at 315°C)	Sand casting	55	50 (EL)	[1]
K1A-F	Investment casting	60	20 (EL)	[3]
AZ63A-T6 (tested at 260°C)	Sand casting	61	15 (EL)	[1]
AM100A-T6 (tested at 150°C)	Sand casting	62	4 (EL)	[1]
ZK51A-T5 (tested at 260°C)	Sand casting	62	16 (EL)	[1]
EZ33A-T5 (tested at 260°C)	Sand casting	69	31 (EL)	[1]
ZE41A-T5 (tested at 315°C)	Sand casting	69	45 (EL)	[1]
AZ81A-T4 (tested at 260°C)	Sand casting	72	35 (EL)	[1]
AZ81A-T4 (tested at 200°C)	Sand casting	76	29 (EL)	[1]
EZ33A-T5 (tested at 200°C)	Sand casting	76	20 (EL)	[1]
AZ81A-T4 (tested at 150°C)	Sand casting	80	24.5 (EL)	[1]
AZ63A-T6 (tested at 200°C)	Sand casting	83	17 (EL)	[1]
AZ81A-T4 (tested at 93°C)	Sand casting	83	20 (EL)	[1]
AZ91C-T6 (tested at 200°C)	Sand casting	83	40 (EL)	[1]
AZ81A-T4	Sand & permanent mould casting	85	15 (EL)	[3]
AZ91E-T4	Sand & permanent mould casting	85	14 (EL)	[3]
ZE41A-T5 (tested at 260°C)	Sand casting	88	40 (EL)	[1]
AM20	Die casting	90	20 (EL)	[5]
ZK51A-T5 (tested at 205°C)	Sand casting	90	17 (EL)	[1]
AZ91E-F	Sand & permanent mould casting	95	3 (EL)	[3]
AZ91C-T6 (tested at 150°C)	Sand casting	97	40 (EL)	[1]
EZ33A-T5 (tested at 150°C)	Sand casting	97	10 (EL)	[1]
Pure Mg	DMD	97 ± 2	7.4 ± 0.2	[6]
AZ81A-T4	Investment casting	100	12 (EL)	[3]
AZ91E-F	Investment casting	100	2 (EL)	[3]
AZ91E-T4	Investment casting	100	12 (EL)	[3]
AZ91E-T5	Investment casting	100	3 (EL)	[3]

EL, elongation (%); DMD, disintegrated melt deposition.

(b) Composite Materials

Material	Processing	Reinforcement	0.2% Yield Strength (MPa)	Ductility (%)	Reference
Mg4Zn/1.5 wt% SiC	Melt casting— ultrasonic cavitation	SiC (50 nm)	72	20.0	[2]
Mg6Zn/1.5 wt% SiC	Melt casting— ultrasonic cavitation	SiC (50 nm)	76	8 (EL)	[4]

7.2. 0.2% YIELD STRENGTH 100–150 MPA AND DUCTILITY MATRIX

(a) Monolithic/Unreinforced Materials

Material	Processing	0.2% Yield Strength (MPa)	Ductility (%)	Reference
AJ52X (tested at 175°C)	Die casting	100	18 (EL)	[3]
AJ62X (tested at 175°C)	Die casting	103	19 (EL)	[3]
AZ63A-T6 (tested at 150°C)	Sand casting	103	15 (EL)	[1]
EZ33A-T5	Sand & permanent mould casting	105	3 (EL)	[1]
EZ33A-T5	Investment casting	110	4 (EL)	[3]
ZRE1 (EZ33)	Casting	110	3 (EL)	[7]
AZ63A-T6 (tested at 120°C)	Sand casting	114	11 (EL)	[1]
ZE41A-T5 (tested at 200°C)	Sand casting	114	31 (EL)	[1]
ZK51A-T5 (tested at 150°C)	Sand casting	115	14 (EL)	[1]
Pure Mg	PM-MS	116.6 ± 11.1	9.0 ± 0.3	[8]
AZ63A-T6 (tested at 93°C)	Sand casting	119	11 (EL)	[1]
AS21	Die casting	120	13 (EL)	[5]
Pure Mg	PM-MS	125 ± 15	5.8 ± 0.9	[9]
AM50	Die casting	125	13 (EL)	[5]
Mg6Zn-T5	Melt casting—ultrasonic cavitation	125	3.5 (EL)	[4]
Pure Mg	DMD	126 ± 7	8 ± 2	[10]
Pure Mg	PM-CS	127 ± 5	9 ± 2	[11, 12]
AM60	Die casting	130	13 (EL)	[5]
ZE41A-T5 (tested at 151°C)	Sand casting	130	12 (EL)	[1]
ACM522 (tested at 175°C)	Die casting	132	9 (EL)	[3]
Pure Mg	PM-CS	132 ± 7	4.2 ± 0.1	[13]
Pure Mg	PM-MS	134 ± 7	7.5 ± 2.5	[14]
Pure Mg	Melt stir	135	12 (EL)	[15]

Material	Processing	0.2% Yield Strength (MPa)	Ductility (%)	Reference
ZE41A-T5 (tested at 93°C)	Sand casting	138	8 (EL)	[1]
AS41	Die casting	140	15 (EL)	[5]
AZ91E-T7	Investment casting	140	5 (EL)	[3]
AJ62X	Die casting	143	7 (EL)	[3]
AE42	Die casting	145	11 (EL)	[5]
ZK51A-T5 (tested at 95°C)	Sand casting	145	12 (EL)	[1]
RZ5 (ZE41)	Casting	148	4.5 (EL)	[7]

DMD, disintegrated melt deposition; PM-MS, powder metallurgy—microwave sintering; PM-CS, powder metallurgy—conventional sintering; EL, elongation (%).

(b) Composite Materials

Material	Processing	Reinforcement	0.2% Yield Strength (MPa)	Ductility (%)	Reference
Mg/0.079Al (volume fraction of Al)	DMD	Al lumps (99.5% purity)	103 ± 10	14.9 ± 2.9 (FS)	[16]
Mg/7.6 wt% SiC	Conventional casting	SiC (25 μm)	112 ± 6	6.8 ± 0.1	[17]
Mg/14.9 wt% SiC	Conventional casting	SiC (25 μm)	112 ± 2	4.7 ± 0.3	[17]
Mg/1 wt% CNT	PM-MS	CNT (40–70 nm) OD	113 ± 3	1.9 ± 0.9 (FS)	[18]
Mg/26.0 wt% SiC	Conventional casting	SiC (25 μm)	114 ± 4	1.8 ± 0.7	[17]
Mg/1 wt% CNT	PM-MS	SiC (50 nm) CNT (40–70 nm) OD	117 ± 6	1.5 ± 0.3 (FS)	[19]
Mg/0.7 wt% Al$_2$O$_3$	PM-MS	Al$_2$O$_3$ (50 nm)	119 ± 7	7.5 ± 0.2 (FS)	[20]
Mg/2.0 wt% Mo	DMD	Mo (44 μm)	119 ± 5	6.4 ± 0.4	[21]
Mg/11.5 wt% SiC-350E	DMD	SiC (38 μm)	119 ± 3	3.9 ± 0.2	[22]
Mg/16.0 wt% SiC	DMD	SiC (25 μm)	120 ± 5	4.7 ± 1.3	[23, 24]
Mg/1.6 wt% CNT	DMD	CNT (20 nm) OD	121 ± 5	12.2 ± 1.7	[11, 25]
Mg/2.0 wt% CNT	DMD	CNT (20 nm) OD	122 ± 7	7.7 ± 1.0	[11, 25]
Mg/0.7 wt% Mo	DMD	Mo (44 μm)	123 ± 15	6.4 ± 1.1	[21]
Mg/3.6 wt% Mo	DMD	Mo (44 μm)	123 ± 4	9.0 ± 2.1	[21]
Mg/11.4 wt% SiC-HT10	DMD	SiC (38 μm)	126 ± 8	10.4 ± 1.6	[26]

Material	Processing	Reinforcement	0.2% Yield Strength (MPa)	Ductility (%)	Reference
Mg/10.3 wt% SiC	DMD	SiC (25 μm)	127 ± 7	6.0 ± 2.3	[23,24]
Mg/0.3 wt% CNT	DMD	CNT (20 nm) OD	128 ± 6	12.7 ± 2.0	[11,25]
Mg/21.3 wt% SiC	DMD	SiC (25 μm)	128 ± 2	1.4 ± 0.1	[23,24]
Mg/7.0 vol.% Al	PM-MS	Al (7–15 μm)	130.2 ± 10.8	4.0 ± 0.9 (FS)	[27]
Mg/1.4 wt% Al$_2$O$_3$	PM-MS	Al$_2$O$_3$ (50 nm)	130 ± 5	7.4 ± 0.3 (FS)	[20]
Mg/11.4 wt% SiC-HT5	DMD	SiC (38 μm)	130 ± 12	12.8 ± 1.5	[26]
Mg/2.5 wt% Al$_2$O$_3$	PM-MS	Al$_2$O$_3$ (0.3 μm)	130.3 ± 3.7	3.9 ± 0.1	[8]
Mg/0.7 wt% CNT-0.3 wt% Al$_2$O$_3$	PM-MS	Al$_2$O$_3$ (50 nm) CNT (40–70 nm) OD	131 ± 6	2.6 ± 1.3 (FS)	[18]
Mg/0.65 wt% SiC	PM-MS	SiC (45–55 nm)	132 ± 14	6.3 ± 1.0	[9]
Mg/1.0 vol.% SiC	PM-MS	SiC (50 nm)	131 ± 12	5.0 ± 0.5 (FS)	[28]
Mg/0.06 wt% CNT	PM-CS	CNT (20 nm) OD	133 ± 2	12 ± 1	[11,12]
Mg/0.5 wt% CNT-0.5 wt% Al$_2$O$_3$	PM-MS	Al$_2$O$_3$ (50 nm) CNT (40–70 nm) OD	137 ± 6	2.5 ± 0.4 (FS)	[18]
Mg/0.18 wt% CNT	PM-CS	CNT (20 nm) OD	138 ± 4	11 ± 1	[11,12]
Mg/11.5 wt% SiC-250E	DMD	SiC (38 μm)	139 ± 7	3.7 ± 0.7	[22]
Mg/1.3 wt% CNT	DMD	CNT (20 nm) OD	140 ± 2	13.5 ± 2.7	[11,12]
Mg/10 vol.% SiC	PM-MS	SiC (25 μm)	140 ± 2	1.5 ± 0.8	[29]
Mg/0.7 wt% CNT-0.3 wt% SiC	PM-MS	SiC (50 nm) CNT (40–70 nm) OD	140 ± 7	2.1 ± 0.5 (FS)	[19]
Mg/0.22 vol.% ZrO$_2$	PM-CS	ZrO$_2$ (29–68 nm)	140 ± 3	6.4 ± 1.5	[30]
Mg/0.5 vol.% Y$_2$O$_3$	DMD	Y$_2$O$_3$ (32–36 nm)	141 ± 7	8.5 ± 1.6	[31]
AZ31B/1.50 vol.% Al$_2$O$_3$	DMD	Al$_2$O$_3$ (50 nm)	144 ± 9	29.5 ± 1.9 (FS)	[32]
Mg/0.92 wt% SiC	PM-MS	SiC (45–55 nm)	144 ± 12	7.0 ± 2.0	[9]
Mg/0.17 vol.% Y$_2$O$_3$	PM-MS	Y$_2$O$_3$ (30–50 nm)	144 ± 2	8.0 ± 2.8	[14]
Mg/0.5 wt% Al$_2$O$_3$	DMD	Al$_2$O$_3$ (50 nm)	146 ± 5	8.0 ± 2.3	[33,34]
Mg/0.30 wt% CNT	PM-CS	CNT (20 nm) OD	146 ± 5	8 ± 1	[11,12]
Mg/1.11 vol.% ZrO$_2$	PM-CS	ZrO$_2$ (29–68 nm)	146 ± 1	10.8 ± 1.3	[30]
AZ31B/1.11 vol.% Al$_2$O$_3$	DMD	Al$_2$O$_3$ (50 nm)	148 ± 11	25.5 ± 2.2 (FS)	[32]
Mg/3.3 wt% Al$_2$O$_3$	PM-MS	Al$_2$O$_3$ (50 nm)	148 ± 10	5.6 ± 0.3 (FS)	[20]
Mg/11.5 wt% SiC-150E	DMD	SiC (38 μm)	148 ± 3	3.6 ± 0.9	[22]

Material	Processing	Reinforcement	0.2% Yield Strength (MPa)	Ductility (%)	Reference
Mg/11.5 wt% SiC-100E	DMD	SiC (38 μm)	148 ± 7	3.6 ± 1.1	[22]
Mg/11.4 wt% SiC-HT0	DMD	SiC (38 μm)	148 ± 7	3.6 ± 1.1	[26]
AZ31B/0.66 vol.% Al$_2$O$_3$	DMD	Al$_2$O$_3$ (50 nm)	149 ± 7	14.6 ± 1.1 (FS)	[32]

DMD, disintegrated melt deposition; PM-MS, powder metallurgy—microwave sintering; PM-CS, powder metallurgy—conventional sintering; HT0, not heat treated; HT5, heat treated for 5 h; HT10, heat treated for 10 h; FS, failure strain (%); OD, outer diameter; E, extrusion temperature (°C).

7.3. 0.2% YIELD STRENGTH 150–200 MPa AND DUCTILITY MATRIX

(a) Monolithic/Unreinforced Materials

Material	Processing	0.2% Yield Strength (MPa)	Elongation (%)	Reference
ACM522	Die casting	158	4	[3]
AZ91	Die casting	160	7	[5]
AJ52X	Die casting	161	13	[3]
Elektron 21	Casting	170	5	[7]
AZ31	DMD	172 ± 15	10.4 ± 3.9 (FS)	[35]
WE43	Casting	180	7	[7]
QE22A-T6	Investment casting	185	4	[3]
WE43A-T6	Sand & permanent mould casting	190	4	[3]
ZE63A-T6	Sand & permanent mould casting	190	7	[3]
AZ91E-T6	Sand & permanent mould casting	195	6	[3]
WE54A-T6	Sand & permanent mould casting	195	4	[3]
EQ21	Casting	195	4	[7]

DMD, disintegrated melt deposition; FS, failure strain (%).

(b) Composite Materials

Material	Processing	Reinforcement	0.2% Yield Strength (MPa)	Ductility (%)	Reference
Mg/0.5 vol.% MgO	DMD	MgO (36 nm)	151 ± 3	8 ± 1	[10]
Mg/1.0 vol.% Y_2O_3	DMD	Y_2O_3 (32–36 nm)	151 ± 5	6.8 ± 0.5	[31]
Mg/1.9 wt% Y_2O_3	PM-CS	Y_2O_3 (29 nm)	151 ± 2	12.0 ± 1.0	[13]
Mg/0.5 wt% CNT-0.5 wt% SiC	PM-MS	SiC (50 nm) CNT (40–70 nm) OD	152 ± 1	2.3 ± 0.6 (FS)	[19]
Mg/0.3 wt% CNT-0.7 wt% SiC	PM-MS	SiC (50 nm) CNT (40–70 nm) OD	153 ± 4	3.3 ± 0.7 (FS)	[19]
Mg/1.11 wt% Y_2O_3	PM-CS	Y_2O_3 (29 nm)	153 ± 3	9.1 ± 0.2	[13]
Mg/2.2 wt% Al_2O_3	PM-MS	Al_2O_3 (50 nm)	154 ± 5	6.3 ± 0.4 (FS)	[20]
Mg/0.3 wt% CNT-0.7 wt% Al_2O_3	PM-MS	Al_2O_3 (50 nm) CNT (40–70 nm) OD	154 ± 2	2.5 ± 0.8	[18]
Mg/9.6 wt% Ti	DMD	Ti ($19 \pm 10 \ \mu m$)	154 ± 10	9.5 ± 0.3	[36]
Mg/15.4 wt% SiC	DMD	SiC ($0.6 \ \mu m$)	155 ± 1	1.4 ± 0.1	[37, 38]
Mg/0.5 vol.% SiC-0.5 vol.% Al_2O_3	PM-MS	SiC (50 nm) Al_2O_3 (50 nm)	156 ± 7	4.6 ± 2.1 (FS)	[28]
Mg/0.6 wt% Y_2O_3	PM-CS	Y_2O_3 (29 nm)	156 ± 1	15.8 ± 0.7	[13]
Mg/1.84 wt% SiC	PM-MS	SiC (45–55 nm)	157 ± 22	7.6 ± 1.5	[9]
Mg/0.7vol.%Y_2O_3	PM-MS	Y_2O_3 (30–50 nm)	157 ± 10	8.6 ± 1.2	[14]
Mg/0.75 vol.% MgO	DMD	MgO (36 nm)	158 ± 5	3 ± 2	[10]
Mg/33.6 wt% SiC	Conventional casting	SiC ($35 \ \mu m$)	158 ± 8	3.1 ± 0.3	[39]
Mg/5.0 wt% Al_2O_3	PM-MS	Al_2O_3 ($0.3 \ \mu m$)	158.5 ± 9.7	2.8 ± 0.3	[8]
Mg/2.0 vol.% Y_2O_3	DMD	Y_2O_3 (32–36 nm)	162 ± 10	7.0 ± 0.5	[31]
Mg/5.6 wt% Ti	DMD	Ti ($19 \pm 10 \ \mu m$)	163 ± 12	11.1 ± 1.4	[36]
Mg/0.66 vol.% ZrO_2	PM-CS	ZrO_2 (29–68 nm)	163 ± 3	11.4 ± 0.9	[30]
Mg/0.3 vol.% SiC-0.7 vol.% Al_2O_3	PM-MS	SiC (50 nm) Al_2O_3 (50 nm)	165 ± 1	4.2 ± 1.8 (FS)	[28]
Mg6Zn-T5	Melt casting— ultrasonic cavitation	SiC (50 nm)	168	3	[4]
AZ31/1.0 vol.% CNT-0.079 AA5052 (volume fraction of Al)	DMD	AA5052 CNT (40–70 nm) OD	169 ± 6	12.2 ± 0.8 (FS)	[40]
Mg/1.0 vol.% MgO	DMD	MgO (36 nm)	169 ± 8	3 ± 1	[10]

Material	Processing	Reinforcement	0.2% Yield Strength (MPa)	Ductility (%)	Reference
Mg/1.5 wt% Al_2O_3	DMD	Al_2O_3 (50 nm)	170 ± 4	12.4 ± 2.1	[33, 34]
Mg/10.2 wt% SiC	DMD	SiC (0.6 μm)	171 ± 3	1.5 ± 0.2	[37, 38]
Mg/2.5 wt% Al_2O_3	PM-CS	Al_2O_3 (1.0 μm)	172 ± 1	16.8 ± 0.4	[41]
Mg/Fe-wire/CF	DMD	Galvanized Fe wire mesh with PAN-based carbon fibers wound around mesh	173 ± 4	3.0 ± 1.4	[42]
Mg/2.5 wt% Al_2O_3	DMD	Al_2O_3 (50 nm)	175 ± 3	14.0 ± 2.4	[33, 34]
Mg/0.7 vol.% Y_2O_3–0.6 vol.% Cu	PM-MS	Y_2O_3 (30–50 nm) Cu (25 nm)	179 ± 7	11.1 ± 0.7 (FS)	[43]
Mg/0.38 wt% Al	PM-CS	Al (18 nm)	181 ± 14	4.8 ± 0.4 (FS)	[44]
Mg/2.5 wt% Al_2O_3	PM-CS	Al_2O_3 (0.3 μm)	182 ± 3	12.1 ± 1.4	[41]
Mg/4.8 wt% SiC	DMD	SiC (0.6 μm)	182 ± 2	2.1 ± 0.9	[37, 38]
Mg/1.52 wt% Al	PM-CS	Al (18 nm)	185 ± 9	3.3 ± 1.0 (FS)	[44]
AZ31B/3.3 wt% Al_2O_3 1 wt% Ca	DMD	Al_2O_3 (50 nm)	185 ± 6	16 ± 1.2 (FS)	[45]
Mg/0.22 vol.% ZrO_2	DMD	ZrO_2 (29–68 nm)	186 ± 2	4.7 ± 0.2	[46]
Mg/1.5 wt% Cu	PM-MS	Cu (50 nm)	188 ± 13	5.9 ± 1.1 (FS)	[47]
AZ31/1.5 vol.% Al_2O_3–0.079 AA5052 (volume fraction of Al)	DMD	AA5052 Al_2O_3 (50 nm)	188 ± 10	11.1 ± 0.4 (FS)	[48]
AZ31/1.0 vol.% CNT	DMD	CNT (40–70 nm) OD	190 ± 13	17.5 ± 2.6 (FS)	[35]
Mg/2.5 wt% Al_2O_3	PM-CS	Al_2O_3 (50 nm)	194 ± 5	6.9 ± 1.0	[41]
Mg/4.9 wt% Cu	PM-MS	Cu (50 nm)	194 ± 17	2.9 ± 0.4 (FS)	[47]

DMD, disintegrated melt deposition; PM-MS, powder metallurgy—microwave sintering; PM-CS, powder metallurgy—conventional sintering; FS, failure strain (%); OD, outer diameter.

7.4. 0.2% YIELD STRENGTH 200–250 MPA AND DUCTILITY MATRIX

(a) Monolithic/Unreinforced Materials

Material	Processing	0.2% Yield Strength (MPa)	Ductility (%)	Reference
AZ91D	PM—mechanical milling	232 ± 6	14 ± 3	[49]

(b) Composite Materials

Material	Processing	Reinforcement	0.2% Yield Strength (MPa)	Ductility (%)	Reference
Mg/1.1 wt% Al$_2$O$_3$	DMD	Al$_2$O$_3$ (0.3 μm)	200 ± 1	8.6 ± 1.1	[6]
AZ31B	DMD	—	201 ± 7	5.6 ± 1.4 (FS)	[32]
MB15	PM-CS	—	202	8.9	[50]
Mg/1.16 wt% Al	PM-CS	Al (18 nm)	202 ± 7	5.0 ± 1.6 (FS)	[44]
AZ31B/1.5 vol.% Al$_2$O$_3$	DMD	Al$_2$O$_3$ (50 nm)	204 ± 8	22.2 ± 2.4 (FS)	[51]
QE22A-T6	Sand & permanent mould casting	—	205	4 (EL)	[3]
MSR-B	Casting	—	205	4 (EL)	[7]
WE54	Casting	—	205	4 (EL)	[7]
Mg/0.7 wt% Al$_2$O$_3$	DMD	Al$_2$O$_3$ (0.3 μm)	214 ± 4	12.5 ± 1.8	[6]
AZ31B/3.3 wt% Al$_2$O$_3$-2 wt% Ca	DMD	Al$_2$O$_3$ (50 nm)	215 ± 5	10 ± 0.2 (FS)	[45]
Mg/0.7 vol.% Y$_2$O$_3$-0.3 vol.% Cu	PM-MS	Y$_2$O$_3$ (30–50 nm) Cu (25 nm)	215 ± 20	11.1 ± 1.0 (FS)	[43]
Mg/1.11 vol.% ZrO$_2$	DMD	ZrO$_2$ (29–68 nm)	216 ± 4	3.0 ± 0.2	[46]
Mg–Al9Zn/15 vol.% SiC	Vacuum stir casting	SiC (12.8 μm)	218	1.1 (EL)	[52]
Mg/0.76 wt% Al	PM-CS	Al (18 nm)	218 ± 16	6.2 ± 0.9 (FS)	[44]
Mg/0.6 wt% Y$_2$O$_3$	DMD	Y$_2$O$_3$ (29 nm)	218 ± 2	12.7 ± 1.3	[53]

Material	Processing	Reinforcement	0.2% Yield Strength (MPa)	Ductility (%)	Reference
Mg/0.7 vol.% Y$_2$O$_3$-0.3 vol.% Ni	PM-MS	Y$_2$O$_3$ (30–50 nm) Ni (20 nm)	221 ± 7	9.0 ± 0.9 (FS)	[54]
Mg/0.66 vol.% ZrO$_2$	DMD	ZrO$_2$ (29–68 nm)	221 ± 5	4.8 ± 0.7	[46]
Mg/2.5 wt% Al$_2$O$_3$	DMD	Al$_2$O$_3$ (0.3 μm)	222 ± 2	4.5 ± 0.5	[6]
Mg-Al9Zn	Vacuum Stir Casting	—	225	7.2 (EL)	[52]
Mg/0.7 vol.% Y$_2$O$_3$-1.0 vol.% Ni	PM-MS	Y$_2$O$_3$ (30–50 nm) Ni (20 nm)	228 ± 8	5.5 ± 0.7 (FS)	[54]
Mg/30 vol.% SiC	Melt Stir	SiC (40 μm)	229	2 (EL)	[15]
Mg/0.7 vol.% Y$_2$O$_3$-0.6 vol.% Ni	PM-MS	Y$_2$O$_3$ (30–50 nm) Ni (20 nm)	232 ± 8	9.5 ± 0.9 (FS)	[54]
AZ31B/3.3 wt% Al$_2$O$_3$-3 wt % Ca	DMD	Al$_2$O$_3$ (50 nm)	235 ± 7	7.3 ± 0.2 (FS)	[45]
Mg/3.0 wt% Cu	PM-MS	Cu (50 nm)	237 ± 24	5.4 ± 1.2 (FS)	[47]

DMD, disintegrated melt deposition; PM-MS, powder metallurgy—microwave sintering; PM-CS, powder metallurgy—conventional sintering; FS, failure strain (%); EL, elongation (%).

7.5. 0.2% YIELD STRENGTH 250–300 MPa AND DUCTILITY MATRIX

(a) Monolithic/Unreinforced Materials

Material	Processing	0.2% Yield Strength (MPa)	Ductility (%)	Reference
AZ91A	DMD	263 ± 12	7 ± 4 (FS)	[55]
AZ91A	DMD (T6)	272 ± 3	3.7 ± 0.5 (FS)	[56]

DMD, disintegrated melt deposition; FS, failure strain (%).

(b) Composite Materials

Material	Processing	Reinforcement	0.2% Yield Strength (MPa)	Ductility (%)	Reference
AZ91D/5%CNT	PM—mechanical milling	CNT (5 μm—length)	277 ± 4	1 ± 0.5	[49]
MB15/10 vol.% Ti-6Al-6V	PM-CS	Ti-6Al-6V (<100 μm)	278	6.0	[50]
AZ91D/0.5%CNT	PM—mechanical milling	CNT (5 μm—length)	281 ± 6	6 ± 2	[49]
Mg/10.1 wt% Cu	DMD	Cu (8–11 μm)	281 ± 13	2.5 ± 0.2	[57,58]
AZ91D/3%CNT	PM—mechanical milling	CNT (5 μm—length)	284 ± 6	3 ± 2	[49]
AZ91D/1%CNT	PM—mechanical milling	CNT (5 μm—length)	295 ± 5	5 ± 2	[49]
AZ91A/15.54 wt% Cu	DMD	Cu (8–11 μm)	299 ± 5	6 ± 1 (FS)	[55]

DMD, disintegrated melt deposition; PM-CS, powder metallurgy—conventional sintering; FS, failure strain (%).

7.6. 0.2% YIELD STRENGTH > 300 MPA AND DUCTILITY MATRIX

(a) Composite Materials

Material	Processing	Reinforcement	0.2% Yield Strength (MPa)	Ductility (%)	Reference
Mg/1.9 wt% Y$_2$O$_3$	DMD	Y$_2$O$_3$ (29 nm)	312 ± 4	6.9 ± 1.6	[53]
Mg/7.3 wt% Ni	DMD	Ni (29 ± 19 μm)	337 ± 15	4.8 ± 1.4	[59]
Mg/18.0 wt% Cu	DMD	Cu (8–11 μm)	355 ± 11	1.5 ± 0.3	[57,58]
AZ91A/17.2 wt% Cu	DMD (T6)	Cu (8–11 μm)	355 ± 8	2.2 ± 0.9 (FS)	[56]
Mg/14.0 wt% Ni	DMD	Ni (29 ± 19 μm)	420 ± 27	1.4 ± 0.1	[59]

DMD, disintegrated melt deposition; FS, failure strain (%).

REFERENCES

1. M. M. Avedesian and H. Baker (eds) (1999) *ASM Specialty Handbook—Magnesium and Magnesium Alloys*. Materials Park, OH: ASM International.

2. G. Cao, J. Kobliska, H. H. Konishi, and X. Li (2008) Tensile properties and microstructure of SiC nanoparticle-reinforced Mg-4Zn alloy fabricated by ultrasonic cavitation-based solidification processing. *Metallurgical and Materials Transactions A*, **39A**, 880–886.

3. R. E. Brown (2006) Magnesium and Its Alloy. In M. Kutz (ed) *Mechanical Engineers' Handbook: Materials and Mechanical Design*, New York, Wiley. vol. **1**, 3rd Edition. John Wiley & Sons, Inc., pp. 278–286.

4. G. Cao, H. Choi, H. Konishi, S. Kou, R. Lakes, and X. Li (2008) Mg–6Zn/1.5%SiC nanocomposites fabricated by ultrasonic cavitation-based solidification processing. *Journal of Materials Science*, **43**, 5521–5526.

5. H. E. Friedrich and B. L. Mordike (eds) (2006) *Magnesium Technology—Metallurgy, Design Data, Applications*. Berlin: Springer, **p. 93**.

6. S. F. Hassan and M. Gupta (2008) Effect of submicron size Al_2O_3 particulates on microstructural and tensile properties of elemental Mg. *Journal of Alloys and Compounds*, **457**, 244–250.

7. Magnesium Elektron. Website: http://www.magnesium-elektron.com (last accessed on June 10, 2008).

8. W. L. E. Wong, S. Karthik, and M. Gupta (2005) Development of high performance Mg-Al_2O_3 composites containing Al_2O_3 in submicron length scale using microwave assisted rapid sintering. *Materials Science and Technology*, **21** (9), 1063–1070.

9. W. L. E. Wong and M. Gupta (2006) Simultaneously improving strength and ductility of magnesium using nano-size SiC particulates and microwaves. *Advanced Engineering Materials*, **8** (8), 735–740.

10. C. S. Goh, M. Gupta, J. Wei, and L. C. Lee (2007) Characterization of high performance Mg/MgO nanocomposites. *Journal of Composite Materials*, **41**(19), 2325–2335.

11. C. S. Goh, J. Wei, L. C. Lee, and M. Gupta (2006) Effect of fabrication techniques on the properties of carbon nanotubes reinforced magnesium. *Solid State Phenomena*, **111**, 179–182.

12. C. S. Goh, J. Wei, L. C. Lee, and M. Gupta (2006) Development of novel carbon nanotube reinforced magnesium nanocomposites using the powder metallurgy technique. *Nanotechnology*, **17**, 7–12.

13. S. F. Hassan and M. Gupta (2007) Development and characterization of ductile Mg/Y_2O_3 nanocomposites. *Transactions of the ASME*, **129**, 462–467.

14. K. S. Tun and M. Gupta (2007) Improving mechanical properties of magnesium using nano-yttria reinforcement and microwave assisted powder metallurgy method. *Composites Science and Technology*, **67**(13), 2657–2664.

15. R. A. Saravanan and M. K. Surappa (2000) Fabrication and characterisation of pure magnesium-30 vol.% SiC$_P$ particle composite. *Materials Science and Engineering A*, **276**, 108–116.

16. M. Paramsothy, N. Srikanth, and M. Gupta (2008) Solidification processed Mg/Al bimetal macrocomposite: microstructure and mechanical properties. *Journal of Alloys and Compounds*, **461**(1–2), 200–208.

17. S. C. V. Lim, M. Gupta, and L. Lu (2001) Processing, microstructure, and properties of Mg–SiC composites synthesised using fluxless casting process. *Materials Science and Technology*, **17**(7), 823–832.

18. S. K. Thakur, T. S. Srivatsan, and M. Gupta (2007) Synthesis and mechanical behavior of carbon nanotube–magnesium composites hybridized with nanoparticles of alumina. *Materials Science and Engineering A*, **466**(1–2), 32–37.

19. S. K. Thakur, T. K. Gan, and M. Gupta (2007) Development and characterization of magnesium composites containing nano-sized silicon carbide and carbon nanotubes as hybrid reinforcements. *Journal of Materials Science*, **42**, 10040–10046.

20. W. L. E. Wong and M. Gupta (2007) Improving overall mechanical performance of magnesium using nano-alumina reinforcement and energy efficient microwave assisted processing route. *Advanced Engineering Materials*, **9**(10), 902–909.

21. W. L. E. Wong and M. Gupta (2005) Enhancing thermal stability, modulus and ductility of magnesium using molybdenum as reinforcement. *Advanced Engineering Materials*, **7**(4), 250–256.

22. S. C. V. Lim and M. Gupta (2001) Enhancing the microstructural and mechanical response of a Mg/SiC formulation by the method of reducing extrusion temperature. *Materials Research Bulletin*, **36**(15), 2627–2636.

23. M. Gupta, M. O. Lai, and D. Saravanaranganathan (2000) Synthesis, microstructure and properties characterization of disintegrated melt deposited Mg/SiC composites. *Journal of Materials Science*, **35**(9), 2155–2165.

24. M. Manoharan, S. C. V. Lim, and M. Gupta (2002) Application of a model for the work hardening behavior to Mg/SiC composites synthesized using a fluxless casting process. *Materials Science and Engineering A*, **333**(1–2), 243–249.

25. C. S. Goh, J. Wei, L. C. Lee, and M. Gupta (2006) Simultaneous enhancement in strength and ductility by reinforcing magnesium with carbon nanotubes. *Materials Science and Engineering A*, **423**, 153–156.

26. S. C. V. Lim and M. Gupta (2003) Enhancing modulus and ductility of Mg/SiC composite through judicious selection of extrusion temperature and heat treatment. *Materials Science and Technology*, **19**, 803–808.

27. S. K. Thakur, M. Paramsothy, and M. Gupta (2010) Improving tensile and compressive strengths of magnesium by blending it with aluminium. *Materials Science and Technology*, **26**(1), 115–120.

28. S. K. Thakur, K. Balasubramanian, and M. Gupta (2007) Microwave synthesis and characterization of magnesium based composites containing nanosized SiC and hybrid (SiC +Al$_2$O$_3$) reinforcements. *Transactions of the ASME*, **129**, 194–199.

29. W.L. E. Wong and M. Gupta (2006) Effect of hybrid length scales (micro + nano) of SiC reinforcement on the properties of magnesium. *Solid state phenomena*, **111**, 91–94.

30. S. F. Hassan, M. J. Tan, and M. Gupta (2007) Development of nano-ZrO_2 reinforced magnesium nanocomposites with significantly improved ductility. *Materials Science and Technology*, **23**(11), 1309–1312.

31. C. S. Goh, J. Wei, L. C. Lee, and M. Gupta (2007) Properties and deformation behavior of Mg-Y_2O_3 nanocomposites. *Acta Materialia*, **55**(15), 5115–5121.

32. Q. B. Nguyen and M. Gupta (2008) Increasing significantly the failure strain and work of fracture of solidification processed AZ31B using nano-Al_2O_3 particulates. *Journal of Alloys and Compounds*, **459**, 244–250.

33. S. F. Hassan and M. Gupta (2004) Development of high-performance magnesium nano-composites using solidification processing route. *Materials Science and Technology*, **20**, 1383–1388.

34. S. F. Hassan and M. Gupta (2006) Effect of type of primary processing on the microstructure, CTE and mechanical properties of magnesium/alumina nanocomposites. *Composite Structures*, **72**, 19–26.

35. M. Paramsothy, S. F. Hassan, N. Srikanth, and M. Gupta (2010) Simultaneous enhancement of tensile/compressive strength and ductility of magnesium alloy AZ31 using carbon nanotubes. *Journal of Nanoscience and Nanotechnology*, **10**(2), 956–964.

36. S. F. Hassan and M. Gupta (2002) Development of ductile magnesium composite materials using titanium as reinforcement. *Journal of Alloys and Compounds*, **345**, 246–251.

37. S. U. Reddy, N. Srikanth, M. Gupta, and S. K. Sinha (2004) Enhancing the properties of magnesium using SiC particulates in sub-micron length scale. *Advanced engineering materials*, **6**(12), 957–964.

38. S. Ugandhar, M. Gupta, and S. K. Sinha (2006) Enhancing strength and ductility of Mg/SiC composites using recrystallization heat treatment. *Composite Structures*, **72**(2), 266–272.

39. M. Gupta, L. Lu, M. O. Lai, and K. H. Lee (1999) Microstructure and mechanical properties of elemental and reinforced magnesium synthesized using a fluxless liquid-phase process. *Materials Research Bulletin*, **34**(8), 1201–1214.

40. M. Paramsothy, S. F. Hassan, N. Srikanth, and M. Gupta (2009) Adding carbon nanotubes and integrating with AA5052 aluminium alloy core to simultaneously enhance stiffness, strength and failure strain of AZ31 magnesium alloy. *Composites Part A: Applied Science and Manufacturing*, **40**(9), 1490–1500.

41. S. F. Hassan and M. Gupta (2006) Effect of length scale of Al_2O_3 particulates on microstructural and tensile properties of elemental Mg. *Materials Science and Engineering A*, **425**(1–2), 22–27.

42. W. L. E. Wong and M. Gupta (2005) Using hybrid reinforcement methodology to enhance overall mechanical performance of pure magnesium. *Journal of Materials Science*, **40**, 2875–2882.

43. K. S. Tun and M. Gupta (2010) Investigating influence of hybrid (yttria+copper) nanoparticulate reinforcements on microstructural development and tensile response of magnesium. *Materials Science and Technology*, **26**(1), 87–94.

44. X. L. Zhong, W. L. E. Wong, and M. Gupta (2007) Enhancing strength and ductility of magnesium by integrating it with aluminum nanoparticles. *Acta Materialia*, **55**(18), 6338–6344.

45. Q. B. Nguyen and M. Gupta (2009) Microstructure and mechanical characteristics of AZ31B/Al_2O_3 nanocomposite with addition of Ca. *Journal of Composite Materials*, **43**(1), 5–17.

46. S. F. Hassan and M. Gupta (2007) Effect of Nano-ZrO_2 particulates reinforcement on microstructure and mechanical behavior of solidification processed elemental Mg. *Journal of Composite Materials*, **41**(21), 2533–2543.

47. W. L. E. Wong and M. Gupta (2007) Development of Mg/Cu nanocomposites using microwave assisted rapid sintering. *Composites Science and Technology*, **67**(7–8), 1541–1552.

48. M. Paramsothy, S. F. Hassan, N. Srikanth, and M. Gupta (2009) Simultaneously enhanced tensile and compressive response of AZ31–NanoAl_2O_3–AA5052 macro-composite. *Journal of Materials Science*, **44**(18), 4860–4873.

49. Y. Shimizu, S. Miki, T. Soga, I. Itoh, H. Todoroki, T. Hosono, K. Sakaki, T. Hayashi, Y. A. Kim, M. Endo, S. Morimoto, and A. Koide (2008) Multi-walled carbon nanotube-reinforced magnesium alloy composites. *Scripta Materialia*, **58**, 267–270.

50. Y. L. Xi, D. L. Chai, W. X. Zhang, and J. E. Zhou (2005) Ti–6Al–4V particle reinforced magnesium matrix composite by powder metallurgy. *Materials Letters*, **59**, 1831–1835.

51. M. Paramsothy, S. F. Hassan, N. Srikanth, and M. Gupta (2009) Enhancing tensile/compressive response of magnesium alloy AZ31 by integrating with Al_2O_3 nanoparticles. *Materials Science and Engineering A*, **527**(1–2), 162–168.

52. M. C. Gui, J. M. Han, and P. Y. Li (2004) Microstructure and mechanical properties of Mg–Al9Zn/SiCp composite produced by vacuum stir casting process. *Materials Science and Technology*, **20**, 765–771.

53. S. F. Hassan and M. Gupta (2007) Development of nano-Y_2O_3 containing magnesium nanocomposites using solidification processing. *Journal of Alloys and Compounds*, **429**, 176–183.

54. K. S. Tun and M. Gupta (2009) Development of magnesium/(yttria + nickel) hybrid nanocomposites using hybrid microwave sintering: microstructure and tensile properties. *Journal of Alloys and Compounds*, **487**(1–2), 76–82.

55. K. F. Ho, M. Gupta, and T. S. Srivatsan (2004) The mechanical behavior of magnesium alloy AZ91 reinforced with fine copper particulates. *Materials Science and Engineering A*, **369**(1–2), 302–308.

56. S. F. Hassan, K. F. Ho, and M. Gupta (2004) Increasing elastic modulus, strength and CTE of AZ91 by reinforcing pure magnesium with elemental copper. *Materials Letters*, **58**(16), 2143–2146.

57. S. F. Hassan and M. Gupta (2002) Development of a novel magnesium–copper based composite with improved mechanical properties. *Materials Research Bulletin*, **37**(2), 377–389.

58. S. F. Hassan and M. Gupta (2003) Development of high strength magnesium–copper based hybrid composites with enhanced tensile properties. *Materials Science and Technology*, **19**, 253–259.

59. S. F. Hassan and M. Gupta (2002) Development of high strength magnesium based composites using elemental nickel particulates as reinforcement. *Journal of Materials Science*, **37**, 2467–2474.

APPENDIX: LIST OF SOME MAGNESIUM SUPPLIERS

Company	Website
Alfa Aesar	www.alfa.com
All Metal Sales, Inc.	www.allmetalsalesinc.com
All Metals & Forge	www.steelforge.com
Allied Metal Company	www.alliedmetalcompany.com
Aluminum Resources, Inc.	www.aluminumresources.com
Belmont Metals, Inc.	www.belmontmetals.com
DCM DECOmetal GmbH	www.dcm-vienna.com
ECKA Granules	www.ecka-granules.com
Electronic Space Products International (ESPI)	www.espi-metals.com
Falcon Metals Group	www.falcon-metals.com
Goodfellow Corporation	www.goodfellow.com
Magnesium Elektron	www.magnesium-elektron.com
Merck	www.merck.com
Micron Metals, Inc.—AEE	www.micronmetals.com
Midwest Metals, Inc.	www.midwestmetalsinc.com
Nippon Kinzoku Co., Ltd	www.nipponkinzoku.co.jp
Reade Advanced Materials	www.reade.com
Sigma-Aldrich	www.sigmaaldrich.com
Standford Materials	www.stanfordmaterials.com
Sudamin	www.sudamin.com
Tangshan Weihao Magnesium Powder Co., Ltd	www.cnmgbp.com
Yangquan Metals & Minerals IMP. & EXP. Co., Ltd	www.yqmm.net

Magnesium, Magnesium Alloys, & Magnesium Composites, by Manoj Gupta and Nai Mui Ling, Sharon
© 2010 John Wiley & Sons, Inc.

ABOUT THE AUTHORS

Associate Professor Manoj Gupta
Associate Professor Gupta is currently the Head of Materials Group at the Department of Mechanical Engineering, National University of Singapore, Singapore. He has published more than 250 internationally peer-reviewed journal papers in the area of metal matrix composites.

Dr Nai Mui Ling, Sharon
Dr Nai wrote this book while she was a Research Fellow at the Department of Mechanical Engineering, National University of Singapore, Singapore. Currently, she is working as an Assistant Research Scientist at the Singapore Institute of Manufacturing Technology (SIMTech).

INDEX

A

Acrylonitrile Butadiene Styrene, 3
Airbag housing, 7
Alkaline electrolyte, 223
Alloy designations, 42, 43
Aluminate electrolyte, 223
Aluminum, 2, 3, 5, 7, 30, 32, 40–43, 45, 63,
    64, 76, 80, 90, 92, 95, 101, 115, 176, 186,
    187, 191–197, 207, 209, 211–213, 222,
    225, 226
Aluminum hydroxide, 221
Aluminum oxide (Al_2O_3), 91–93, 95,
    115–119, 121–134, 158, 194, 195, 217,
    223
Ammonium bifluoride, 221
Anodic, 208, 209
Anodizing, 217, 219, 220, 221, 223, 224
Applications 5–11
    aerospace, 7
    automotive, 5
    electronics, 10
    medical, 8
    optical, 10
    sports, 9
Archery bow, 9

B

Beryllium, 40
Bicycle frame, 10
Biocompatible, 9
Biodegradable, 9
Blending, 21, 22, 118
Bonding, 94
    chemical, 94
    mechanical, 94
Borate solution, 224
Boron carbide, 115

C

Calcium, 40
Camera housing, 10
Carbon, 115
Carbon nanotube (CNT), 91, 105, 114, 115,
    118–121, 133, 134, 159, 170–176, 196,
    197
Carbon fiber, 3, 217
Cathodic, 208, 209
Cathodic epoxy electrocoating, 226
Cathodic site, 213
Ceramic matrix composite (CMC), 88
Cerium, 40
Chain saw housing, 11
Chemical conversion coating, 217
Chromate solution, 220
Cleavage crack, 120
Clustering tendency, 117, 118
Coefficient of thermal expansion, 99, 100
    Kerner's model, 100
    Rule of mixture (ROM), 99
    Turner's model, 99
Compaction, 22, 24–27
    compaction, cold, 24, 26
    compaction, hot, 24, 25
    isostatic pressing, 24, 25
    isostatic pressing, hot, 25
    isostatic presssing, cold, 25
    uniaxial pressing, 24
Contact angle, 94
Conversion coating, 219
Copper, 40, 43, 90, 115, 176–180, 208, 213,
    215
Corrosion, 207–231
    galvanic, 211
    high temperature oxidation, 211, 214
    intergranular, 211, 212
    localized, 212

Magnesium, Magnesium Alloys, & Magnesium Composites, by Manoj Gupta and Nai Mui Ling, Sharon
© 2010 John Wiley & Sons, Inc.

Printed in the United States
By Bookmasters